Collins

Scottish
Birds

KU-448-567

Collins

Scottish Birds

VALERIE THOM

Valerie Thom worked as an agricultural adviser and in conservation education before becoming a freelance writer on the Scottish countryside. She wrote the Scottish section of the *Macmillan Guide to Britain's Nature Reserves* (1984), *Fair Isle: An Island Saga* (1989) and *Birds in Scotland*, published in 1986 to mark the Golden Jubilee of the Scottish Ornithologists' Club, of which she was an Honorary President.

HarperCollins Publishers
77-85 Fulham Palace Road
London W6 8JB
www.collins.co.uk

First published 1994
This edition reissued 2008

10 9 8 7 6 5 4 3 2

Text © The Scottish Ornithologist Club 1994, 2000
Artwork © HarperCollins*Publishers* 1987, 2000
Photographs © The Scottish Ornithologist Club 1994, 2000,
unless otherwise stated

ISBN 978 0 00 727068 2
All rights reserved

Illustrations by Norman Arlott
Gaelic names kindly provided by Calum Campbell

Thug Comhairle nan Leabhraichean tabhartas barantais
airson gun cuirte ainmean Gàidhlig nan eun dhan leabhar

Also available in this series: *Collins Scottish Wild Flowers*

Printed and Bound by Printing Express Ltd., Hong Kong.

CONTENTS

ABOUT THIS BOOK

This book is intended for anyone who knows relatively little about Scottish birds and would like to know more. It aims to help you, in a user-friendly way, to recognise most of the birds you can expect to see in Scotland. Unlike many other field guides it does not cover birds which visit only occasionally, or which occur in such small numbers or are so difficult to identify that only dedicated and experienced birdwatchers are likely to spot them. Instead it concentrates mainly on the commoner species, plus a few scarcer ones included because they are especially 'Scottish' or are of particular interest for some other reason. As well as highlighting the features which help in identifying a bird, it gives information on the habitat(s) occupied by each species, the season during which it is present, and the areas in which it occurs. It also suggests some good places to visit for watching various groups of birds.

In many field guides the species descriptions are arranged in scientific order, but this is not easy to understand and also means that birds likely to be seen together are widely scattered in the book, making it difficult to find all the 'right' pictures quickly. I have chosen instead to group the species according to the habitat in which they are most likely to be seen. This allows you to use the book in two different ways. You can prepare for a trip to a particular type of countryside by reading the relevant habitat introduction so that you will know what species to look out for. Or you can check quickly through the limited number of pictures relating to the habitat you are in when you see a bird you do not know. The most accessible habitats (Gardens and Farmland) are dealt with first, as familiarity with the birds found there is helpful when trying to spot and identify woodland birds.

THE DESCRIPTIONS OF HABITATS AND THEIR BIRDS

Birds are very mobile creatures and often make use of different habitats at different times of year; it is consequently seldom possible to say that a particular species occurs only in a particular type of habitat. In this book the birds are grouped in the habitats in which they are most likely to be seen or most easily watched. Within each habitat chapter the species are arranged in the sequence that I have felt to be most helpful: in general this means that the larger birds come first, and that confusing species are on facing pages wherever practicable. But no arrangement is perfect, and there are undoubtedly some inconsistencies! Scattered among the species descriptions there are a number of 'boxes' which give information on the characteristics of particular groups of birds, and hopefully will help in the elimination process.

Each species description starts with a very brief summary of status: this indicates (1) whether the bird is WIDESPREAD and occurs in most districts with suitable habitat or only in certain limited LOCAL areas, and (2) whether it is likely to be seen at any time of year or only at certain periods. Some Scottish breeding birds move to a different habitat in winter, or move south and are replaced by other individuals of the same species coming from further north, which means that not all individuals of a species represented throughout the year are actually permanent residents in a particular area. And some species are more likely to be seen when on migration in spring and autumn than at any other time. Helpful identifying features are shown in *italics*. Wherever a chapter or species account refers to a bird not described in that section the relevant page number is given to make cross reference as easy as possible.

SOME GENERAL HINTS ON BIRD IDENTIFICATION

Many different aspects of a bird's appearance and lifestyle can assist in its identification – and also add a great deal to the interest of birdwatching. Some are obvious:

(1) Its size – larger, smaller or similar to a familiar bird, such as a house sparrow, blackbird or woodpigeon.

Relative sizes of some common birds

(2) Its proportions – relative length of neck, legs and tail, stocky or slender body.

(3) The shape of its wings and tail – wings pointed or rounded, tail long or short, forked or rounded.

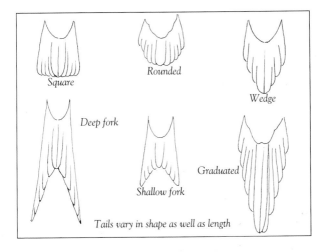

Square

Rounded

Wedge

Deep fork

Shallow fork

Graduated

Tails vary in shape as well as length

(4) The size and shape of its bill – small and slim for picking out insects, short and strong for seed cracking, long and slender for probing in mud, or hooked for tearing flesh.

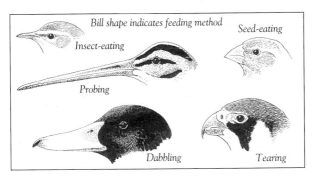

Bill shape indicates feeding method

Insect-eating

Seed-eating

Probing

Dabbling

Tearing

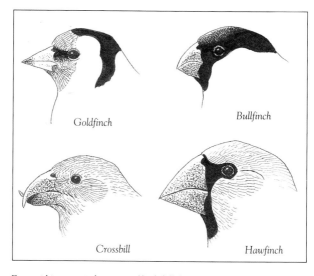

Goldfinch

Bullfinch

Crossbill

Hawfinch

Even within a particular group of birds bill shapes may vary:
Finches – the harder the food, the heavier and stronger the bill
Ducks – bills are adapted to cope with the main food taken (see pp.124-5)

Red-breasted
merganser

Common
scoter

Shoveler

Wigeon

Goldeneye

(5) Its general colour (which often varies with sex and age and sometimes with season) and the position and colour of any distinctive markings – especially patches on wings, rump or tail, stripes on the head or bars on the wings.

White – or coloured – markings are useful aids to identification

Shoulder patch
(chaffinch)

Wing bar
(tufted duck)

White rump
(bullfinch)

White side
to tail
(whinchat)

Speculum
(mallard)

White outer
tail feathers
(yellowhammer)

Wing patch
(goldeneye)

White tail tip
(mistle thrush)

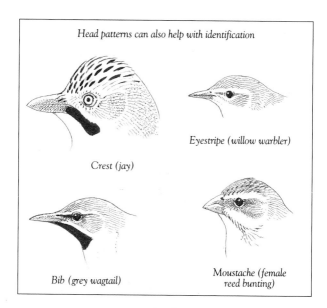

Head patterns can also help with identification

Eyestripe (willow warbler)

Crest (jay)

Bib (grey wagtail)

Moustache (female reed bunting)

Other characteristics are less obvious but often equally useful in confirming a bird's identity:

(6) The way it moves – does it walk, hop, run, crouch, climb trees; flies – fast and direct, in loops, hovering or soaring, returning to the same perch; dives into water – plunging head or feet first from a height, sliding under the surface or giving a quick jump up first; uses its tail – cocking, flicking, fanning or wagging.

(7) The way it feeds – picking things off plants or the ground, catching flies or other birds in flight, pouncing on prey, probing deeply or shallowly.

(8) The season, and the habitat the bird is using – many species are in Scotland for only part of the year and some occur in only very specific types of habitat. And, of course,

(9) the noises it makes – some birds are elusive and difficult to see but can be readily identified by their song or calls – with practice! It is not easy to describe bird noises in words, though this has been attempted for most of the species described. Such efforts are of limited value, however, and much better ways of learning to recognise calls and songs are to go out with an experienced birdwatcher and to listen to good recordings. It helps to start in early spring, so that you are familiar with the resident species' songs before the summer visitors arrive.

A CODE OF CONDUCT FOR BIRDWATCHERS

Today birdwatchers are more numerous than ever before – and birds are under greater pressures. Some ways in which everyone who is interested in birds can help in conserving them, and the habitats upon which they depend, are listed below.

Help the birds by keeping disturbance to a minimum
- avoid scaring birds at the nest because this increases opportunities for predators to take eggs or young
- take extra care in places where ground nesting birds breed, especially in colonies; well camouflaged eggs can easily be unnoticed and trodden on
- make sure that you do not stay long near a nest; to do so can result in chilling of eggs or young
- take care not to disturb feeding birds unnecessarily in very cold weather; they may be struggling to survive

- remember that rare breeding birds are at special risk from egg-collectors; visit known haunts only if they are adequately protected

Help to conserve birds by protecting their habitats
- take care not to start fire on heaths or in woodlands
- keep to paths wherever possible
- remember that some forms of litter can be lethal to birds

Respect the law and other people's rights
- abide by the bird protection laws at all time (copies of *Wild Birds and the Law* can be obtained from the RSPB)
- always obey the Country Code
- respect the rights of landowners; always get permission before entering on private property

(Based on the code produced by the RSPB, The Lodge, Sandy, Bedfordshire SG19 2DL)

GOING FURTHER WITH BIRDWATCHING

Many people who first become interested in birds by watching them at a bird table later find that they want to discover much more about them than can be included in a pocket guide like this. In the Appendix you will find the names and addresses of some organisations which can be of help in this connection, and also some suggestions for books which will assist you in widening your knowledge of Scotland's birds. And a pair of binoculars is really essential for birdwatching satisfaction.

GARDENS, PARKS AND BUILDINGS

Strange-looking birds, not shown in any book, sometimes turn up in a garden (see p. 27)

(see p. 27)

Some bird species have become accustomed, over the years, to living close to man, and taking advantage of the benefits that may come from this association. Such benefits include the regular supply of suitable food – and water – put out specially for birds; the shelter and natural food obtainable on, under or around cultivated trees, shrubs and other plants; and the varied nesting – and roosting – sites provided by buildings. Living close to man is not always beneficial, however. Winter food supplies may be plentiful but there can be a shortage of the caterpillars and other insects needed for rearing young, while cats are much more of a threat in gardens and parks than elsewhere.

Birds which live in close association with man often pay little attention to the regular comings and goings of nearby people, and sometimes

Many of the birds found in parks are accustomed to the presence of people and are often quite tame

become very tame indeed. Some blackbirds and tits even become so accustomed to being fed that they start to draw attention to their needs if feeding-time is late, flying up to a window, tapping at the glass, or simply screeching outside until food appears.

GARDENS AND PARKS

Most people start their birdwatching in a garden or park, where many of the birds are tame enough to give good close-up views. The majority of garden birds were originally inhabitants of woods or scrub and are still found in these habitats too, so familiarity with them at close range can be a big help when trying to identify birds seen further afield. Garden birdwatching also gives good opportunities for studying such things as plumage pattern, bill shape, feeding method, perching position, and flying style, all of which can be useful pointers in identification.

The number and variety of birds visiting a particular garden or park is determined more by the nature of the plant cover and the food supply than by size, although for some species this is a factor. Some birds need tall trees to sing from or roost in, while others find all they need in the way of shelter among shrubs. Some require rather dense bushy growth at ground level, where they can search for insects without going far from cover, while others prefer to feed on open ground. And some come for

Even quite a small pond will attract many different species; the fence round this one is to discourage herons from wading in to catch goldfish!

very specific kinds of food, such as berries or fruit buds, while others take almost anything they can find. In general, gardens with a mixture of trees, shrubs and open ground, or which form part of a closely linked complex of gardens providing these features, will support the greatest variety of birds – and the presence of a pond and of berry-bearing shrubs are definite attractions. The less well-managed the garden the better, from the birds' point of view if not the gardener's, as both insect pests and weed seeds are readily eaten by many garden birds.

The 'top ten' garden birds – those most often taking food and water put out for them – according to the British Trust for Ornithology's Garden Bird Feeding Survey, are:

1. Blue tit	6. Chaffinch
2. Robin	7. Greenfinch
3. Blackbird	8. Great tit
4. House sparrow	9. Dunnock
5. Starling	10. Coal tit

All these species come for seed, fat or scraps at bird tables or to nut baskets, although dunnocks prefer to forage on the ground below rather than visit the table itself. Blue tits, chaffinches and house sparrows are so tolerant of man's proximity that they will readily feed at seed or nut containers fastened onto a window pane, or on a windowsill, so not even a garden is necessary in order to enjoy watching birds. Wren, song thrush and collared dove are also regular garden birds, the first two normally finding their own food while loose peanuts will attract the doves.

Bread, chicken skin and similar scraps are often seized upon by black-headed gulls (p.143) and in some areas herring gulls (p.191), while jackdaws

A well-stocked bird table will bring birds close to a window, where they can be easily watched. Photo by John Hawkins

sometimes greedily stuff themselves with peanuts. Less frequent visitors include bullfinches, which come most often in spring to feed on the buds of fruit and flowering cherry trees; magpies, in districts where this species is abundant; and sparrowhawks (p.86), which find easy prey among the small birds congregated around a bird table. Parties of long-tailed tits (p.108) occasionally come to nut baskets, and overwintering blackcaps (p.101) are increasingly often seen. Hawfinches favour really large gardens and parks but can be very difficult to locate, whereas waxwings, irregular winter visitors, sit around conspicuously on berry-bearing shrubs.

Hard winter weather, especially a combination of frost and snow, brings in a variety of other birds which are finding food hard to come by elsewhere. Redwings and fieldfares (p.71) come for berries, great

The berries of cotoneasters and other shrubs provide food for waxwings – if they have not been eaten by thrushes before the waxwings arrive. Photo by Roger Tidman

spotted woodpeckers (p.90) for fat, bramblings (p.75) for seed, siskins (p.113) for peanuts, woodpigeons (p.69) for cabbages and other brassicas, and crows (p.65) for anything that is going. When frost and snow last for more than a few days even such unlikely birds as pheasants (p.59) and moorhens (p.138) may turn up in some gardens.

BUILDINGS AND OTHER MAN-MADE STRUCTURES

Several bird species select buildings as their preferred nest site, and would now find it difficult to survive without such ready-made accommodation. These are mainly the hole-nesters, for example starlings, house sparrows, jackdaws and swifts, but also house martins, which originally nested on cliffs, as a few still do. Most swallows nowadays build on the rafters of outbuildings with a permanently open door or window, while feral pigeons occupy a wide variety of nooks and ledges, and barn owls (p.68) often nest in semi-derelict buildings. Less usual building-nesters include kittiwakes (p.191), kestrels (p.67) and even a few peregrines (p.165) on the ledges of buildings, and herring gulls and oystercatchers (p.204) on flat roofs. Buildings also provide roosting places, sometimes attracting large numbers of birds in winter when it is important that body heat is maintained. Huge flocks of starlings occupy city centre ledges, benefiting from the warmth given off by street lights, or hurtle low across the carriageway of the Forth Bridge as they gather to roost on the struts. Pied wagtails also make use of buildings, with hundreds finding shelter in derelict hangars and smaller groups going into greenhouses.

Man-made structures such as drystone walls and bridges offer nesting sites for pied wagtails and dippers

Nearly all house martins now nest on buildings, only a few remaining faithful to their natural home on a cliff. Photo by R. Wilmshurst

(p.149), among others. Some types of man-created habitat, though not exactly structures, also attract significant numbers of birds. The most obvious of these are roads and their associated verges. Even alongside busy motorways oystercatchers and lapwings (p.63) can be seen shepherding their young or feeding along the grass verge; overhead kestrels regularly hover in search of mice; and on the roadway itself pied wagtails hawk for flies, while crows and gulls scavenge the remains of any creatures hit by passing vehicles.

Jackdaw

Corvus monedula
Gael: *Cathag*

ADULT

Widespread   **All year**

Jackdaws, the *smallest of the 'black crows'*, are readily identified by the *pale grey* area around the *sides and back of the head*. Their underparts are dark grey, rather than black, and young birds have a brownish tinge. Adults have distinctive pale grey eyes. A fairly sharp, high-pitched 'chak' is the usual call, much used in flight to keep the members of a flock in contact with one another, and run together when the birds are excited to give a chattering 'chaka-chaka-chak'.

The jackdaw is a neat and alert bird, which walks about jauntily, often glides near cliff faces, and performs aerial acrobatics, dodging, twisting and diving as though enjoying it. Its varied diet includes insects, cereals, and peanuts and scraps at bird tables, and it will steal eggs and hide food for later use. Very gregarious by nature, jackdaws form large flocks, often feeding with rooks and starlings, and roosting with rooks, sometimes in huge gatherings. They are catholic in habitat choice, occurring in towns and wooded farmland, and on hillsides and coastal cliffs, and often nest socially, with many pairs occupying adjacent holes. In towns unused chimneys are a favourite site.

ADULT

23

Magpie

Pica pica
Gael: *Pioghaid*

ADULT

Local   **All year**

The magpie's very distinctive *pied plumage* and *long wedge-shaped tail* make it impossible to confuse with any other bird. Its flight is direct and rather slow and the birds often fly in single file. A rapid, harsh chattering 'cak-cak-cak' is the commonest call; the species has no recognisable song.

Magpies are most abundant in the Forth–Clyde valley and along the Aberdeen–Kincardine coastal strip, and are most likely to be seen in parks, around suburban gardens, and on farmland with unkempt hedges and rough grassland. They are very local on low ground elsewhere in Scotland and are absent from the highlands and islands. In the past this species was heavily persecuted by keepers, and it is still regarded as a pest as it eats the eggs and young of game and other birds. Grain, seeds, insects and worms are also taken, as the birds walk and hop around, usually with the tail a bit raised. Rather surprisingly in view of their long tails, magpies build domed nests, usually siting these untidy looking structures high up in a tree or hedge.

ADULT

24

Collared Dove

Streptopelia decaocto
Gael: Calman Coilearach

ADULT

Widespread   **All year**

This elegant small pigeon is very closely associated with man and is a frequent and tame garden visitor. The *white tips to its outer tail feathers* are very obvious in flight or when the bird fans its tail while sunning itself. The *collared* dove's distinctive 'ku-kroo-ku' song, with the accent on the middle syllable, is occasionally mistaken for the call of the cuckoo; it has a harsh, nasal 'kwarr' flight call. It flies fast, with looping falls, glides and dives, bobs its head when walking, and frequently perches on TV aerials, lamp posts and roofs.

Collared doves are now such a familiar sight that it is hard to believe that they first appeared in Scotland as recently as 1957. They have been seen as far north as Shetland, but are commonest around towns, villages and farms in the central lowlands and up the east coast to Dornoch. They have an unusually long breeding season, allowing them to raise several broods a year, which has doubtless helped their rapid spread. Unlike their larger relations, they do little damage to cultivated vegetables although they are vegetarians. They come readily for peanuts.

ADULT

Starling

Sturnus vulgaris
Gael: Druid

ADULT WINTER

ADULT SUMMER

Widespread   **All year**

In winter *buffish spots* give the quarrelsome starling a *speckled appearance*, but these literally wear off as the feathers become worn and by summer have almost vanished. Newly fledged young birds are unspotted mouse-brown with a whitish chin patch. The starling's song is a lively rambling warble, usually including whistles, gurgles and clicks, and it is a good mimic – curlew calls and goose noises often indicate the arrival of these species in the area. Its 'conversation' includes a grating 'tcheerr' and a harsh alarm scream.

Starlings are opportunists and take full advantage of man's activities in many ways, nesting in the roofs and walls of buildings, roosting in vast numbers along wires in city streets and under bridges, and feeding readily at bird tables and on rubbish tips. They feed in the open, bustling about with half-opened bills as they probe the ground for insect larvae – usefully aerating the garden lawn and reducing the number of pests! They are very social birds, both feeding and roosting in large gatherings; the aerial evolutions performed by flocks approaching the roost are fascinating to watch, as they wheel and dive.

JUVENILE

Blackbird

Turdus merula
Gael: Lon-Dubh

ADULT MALE

Widespread   **All year**

The sombrely coloured blackbird is one of the most familiar of garden residents. Young birds and *females are brown* rather than black and lack the adult *male's bright orange bill*. This species not infrequently produces pied partial albinos, which range from black with white spots or collar to almost completely white. Blackbirds are early risers and among the first contributors to the dawn chorus; their song is rich and melodious, with each phrase lasting for several seconds and delivered at a leisurely pace. Alarm calls include a screeching chatter and a loud 'chink-chink' when there is a cat or owl in the vicinity.

Blackbirds occur almost everywhere there are trees, but are relatively scarce in highland birchwoods and in conifer woodlands. They nest, often only a few feet up, in bushes or hedges; the nest is made of twigs, grass, moss and mud, and is lined first with mud and then with dry grass. Two or three broods may be raised. They feed largely on worms, but also take insects, berries and household scraps; individuals sometimes become very tame when regularly fed.

JUVENILE

See 'Thrushes' p.70

Song Thrush

Turdus philomelos
Gael: *Smeòrach*

ADULT

Widespread   **All year**

This *spotty-breasted* species is shyer than the blackbird; it spends much time in or close to cover, is a less frequent visitor to bird tables, and suffers badly in severe winters. Its *varied and musical song,* often delivered from a TV aerial or tree top, is made up of *short phrases, each repeated two or three times,* with pauses between. Its usual call is a thin 'sipp' or 'tick'. When feeding it typically runs or hops for a short distance, then stands with head cocked to one side.

The song thrush is widely distributed wherever there are shrubs or trees, but is much less abundant than the blackbird. Its nest is similar to a blackbird's but without the mud. In addition to worms, insects and berries this species eats snails, which it breaks by hammering the shell on a stone; damaged apples or pears, placed close to bushes, are also readily taken. In autumn many song thrushes move to warmer climes; in a hard winter those which remain here may die of starvation as frozen ground and snow cover prevent them from obtaining enough food.

ADULT

See 'Thrushes' p.70

28

Sparrows & Buntings

House (p.30) & tree (p.72) sparrow, yellowhammer
(p.74), reed (p.151), corn (p.76) & snow (p.212) buntings
- are roughly house sparrow sized and have a similar
  stocky shape
- have short, almost conical, seed-eating bills
- have longish, squared tails
- sexes differ in all species included here except tree
  sparrow
- few have wing bars or patches (snow bunting is an
  exception)

Finches

Chaffinch (p.31), greenfinch (p.32), bullfinch (p.33),
hawfinch (p.34), goldfinch (p.73), brambling (p.75), siskin
(p.113), linnet (p.176), twite (p.176), redpoll (p.112) &
crossbill (p.114)
- vary in size from slightly smaller than a sparrow to
  slightly larger
- have strong bills adapted for dealing with seeds
  of varying size, hardness and accessibility (see p.10)
- have shallowly forked tails
- sexes differ in some species, are alike in others
- the position and colour of wing and tail markings, best
  seen in flight, help in identification:
  white wing bar or patch – chaffinch & hawfinch
  white rump – bullfinch, goldfinch & brambling
  white sides to tail – chaffinch & linnet
  yellow wing bar or patch – greenfinch, goldfinch & siskin
  yellow sides to tail – greenfinch & siskin
- young birds of the following species are streaky
  greenish or brownish and can be difficult to identify:
  greenfinch, goldfinch, siskin, linnet, twite & redpoll.
  (See also juvenile yellowhammer and reed bunting)
  Habitat is often a useful clue to the likely species.

House Sparrow

Passer domesticus
Gael: *Gealbhonn*

MALE

Widespread   **All year**

The bold, yet wary, house sparrow is familiar to everyone who gives birds even a passing glance. The only species with which the male can be confused is the closely related tree sparrow (p.72), which has a chestnut not a *grey crown*. Females and young birds are vaguely similar in appearance to young chaffinches, but are much browner and have shorter *tails, without any white*, and only the faintest suggestion of a wing bar.

This is the species most closely associated with man; it occurs only in the vicinity of buildings and cultivated fields. House sparrows feed mainly on the ground – in the countryside on grain, in towns on scraps, and in gardens on seeds and insects. They can be a nuisance in gardens as they have an annoying habit of eating flowers, especially yellow crocuses and polyanthus, and sometimes descend in flocks to eat the leaves of pea plants. They are busy, bustling birds which indulge in frequent squabbles as they hop around or gather in bushes where they roost socially. This species is gregarious at all times and often joins up with finch flocks in winter.

See 'Sparrows' p.29

FEMALE

Chaffinch

Fringilla coelebs
Gael: Bricein Beithe

MALE SUMMER

Widespread **All year**

Its *double white wing bars* and *white outer tail feathers* help to identify the chaffinch, whatever its age or sex, and distinguish the rather dull-coloured females from female house sparrows. The male's distinctive pinkish-brown breast and blue-grey head are brighter in spring than in winter. A loud 'pink' or 'chwink' and in flight a rather subdued 'tseeip' are the chaffinch's commonest calls; its song is lively and relatively short, accelerating as it descends the scale and ending with a fast up and down flourish – 'cherwit-teeoo'.

The chaffinch is among the commonest British birds and has adapted to life in many different habitats, from town gardens to hillside birchwoods – wherever there are trees and shrubs. It feeds mainly on seeds in winter, but on insects among tree foliage in summer. It is also quick to take advantage of man's left-overs, readily coming to bird tables and picnic places. Scottish chaffinches remain close to their nesting area all year, but in winter immigrant birds from Scandinavia arrive, often in single sex flocks, and join up with other finches to forage widely in search of food.

See 'Finches' p.29

FEMALE

31

Greenfinch

Carduelis chloris
Gael: Glaisean Daraich

MALE

Widespread   **All year**

Bright yellow patches on the wings and sides of the tail distinguish the greenfinch in all plumages. Females are brownish-green, faintly streaked, and young birds brown, heavily streaked. Calls include a long drawn out nasal 'dzeee' and, in flight, a trilling musical 'chi-chi-chi'. Greenfinch song is a rather monotonous warbling twitter, including wheezy 'dzeee' notes, delivered from the top of a tree, or in a slow-flapping song flight.

Greenfinches were until quite recently birds of farmland, nesting in hedges and wintering on the grain and weed seeds around stackyards. Since stacks vanished from the farming scene this species has become much more a garden bird, coming regularly to bird tables. It is partial to peanuts and, although primarily a ground feeder, has learnt to hang on nut baskets. It also likes sunflower seeds, which its bill is strong enough to split open. Ringing has shown that although there may be fewer than ten birds in a garden on any one day, over a winter several hundred probably visit it. Greenfinches are widespread in all lowland areas of Scotland. In winter many join flocks of chaffinches, yellowhammers (p.74) and buntings foraging over the fields.

FEMALE

See 'Finches' p.29

Bullfinch

Pyrrhula pyrrhula
Gael: *Corcan Coille*

MALE

Widespread   **All year**

A *white rump*, very obvious in flight, and a *neat black cap* are conspicuous features of adult bullfinches, which are rather *dumpy*, plump-looking birds. Young birds are pale greyish brown, lack the black cap, and have light coloured bills. Bullfinches do not have a distinctive song, but their soft piping whistle is often the first indication of their presence, as they seldom descend to the ground and prefer to remain in the cover of trees or shrubs.

The bullfinch's diet is mainly tree buds and seeds. In the south this species does a lot of damage in orchards by stripping fruit buds but in Scotland this is not such a problem, though bullfinches visiting gardens in early spring often take the buds of flowering cherries. Their natural habitat is scrubby deciduous woodland but they sometimes frequent conifer plantations at the thicket stage, and feed on heather seeds nearby. Bullfinches are more solitary than other finches and are seldom seen in groups bigger than family parties. They are scarce over much of the northern highlands and absent as a breeding bird from the Northern and Western Isles.

See 'Finches' p.29

FEMALE

Hawfinch

Coccothraustes coccothraustes
Gael: Gobach

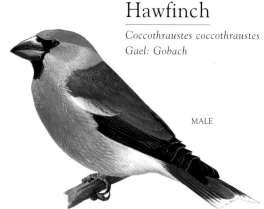

MALE

Local   **All year**

This is an elusive and wary bird, which is most likely to be
seen as it comes into the open to drink or bathe, when its
massive head and bill are very obvious, or as it flies between
trees, when it shows *broad white wing bands* and a *white tip to
its tail*. Its usual call is a brisk 'tik', which often discloses its
presence high in the treetops.

Hawfinches feed on large seeds, such as cherry and
blackthorn, as well as berries such as rose-hips and
hawthorn; their skulls are specially strengthened to support
the very strong muscles which enable them to crack even
cherry stones. This species is scarce
and local in Scotland. The two
places from which it is most
frequently reported are the
Royal Botanic Gardens in
Edinburgh and the grounds of
Scone Palace near Perth, at both of
which there is a wide enough range
of suitable fruit-bearing trees to
ensure an adequate food supply.

FEMALE

See 'Finches' p.29

Waxwing

Bombycilla garrulus
Gael: *Gocan Cireanach*

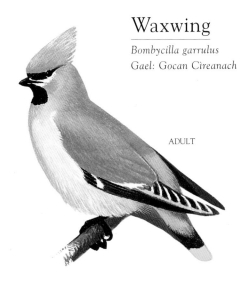

ADULT

Local & irregular   **Between September & April**

The waxwing's *crest* is striking, even if the bird is seen only in silhouette, when its *dumpy shape* and *short tail* are also apparent. In good light the closed wing shows as boldly marked with white, red and yellow towards the tip, and the *yellow tail tip* is obvious. Waxwings in this country tend to be rather silent, but occasionally give a quiet trill, usually in flight.

This is one of the so-called irruptive species which visit this country, irregularly and in varying numbers, because the crop of berries, especially rowans, in their Scandinavian breeding area has proved inadequate to sustain them over the winter. Waxwings are gregarious and usually move around in flocks, the size of which depends upon the scale of the irruption. In some winters several hundred appear; in others none at all. They are most likely to be seen on berry-bearing shrubs, such as cotoneaster, and often stay around the same area for several days, spending many hours just sitting about.

Pied Wagtail

Motacilla alba
Gael: *Breac an t-Sìl*

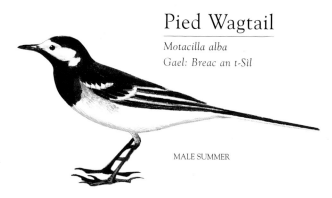

MALE SUMMER

Widespread   **All year**

This is the only *small* ground bird with predominantly *black and white plumage* and a *long tail*. Females and young birds are greyer on the back, and in winter all pied wagtails look markedly whiter in the face, as they lose much of the black bib, only a crescent shaped 'scarf' remaining. A brisk 'chizzik' is the commonest call; it is often given in flight and is quite distinctive. Similar notes are included in the song, which is a lively warbling twitter.

Pied wagtails are widespread in towns and gardens, around farm steadings, and in open country – where they often feed along the roadside – wherever there is a good supply of insect food and of nesting sites, such as holes in walls or buildings, crevices in steep banks, or thick ivy. They are very active feeders, running about picking up insects and often making short fly-catching flights. Between August and March pied wagtails join up in flocks and roost communally, sometimes in flocks of several hundred, in reedbeds or among willows; in severe weather they often move into little-used buildings for added protection.

FEMALE
WINTER

Robin

Erithacus rubecula
Gael: *Am Brù-Dhearg*

ADULT

Widespread   **All year**

Although by no means the commonest British bird the 'robin redbreast' is probably the best-known. Young birds lack the *red breast* and are a scaly-looking golden brown; by winter they are indistinguishable from adults. Both sexes sing, and only for a short period in mid-summer does song stop altogether. The song, a short and melodious warble, has a plaintive quality which is distinctive and easy to recognise. The commonest calls are a sharp, scolding 'tik, tik', often used when birds are alarmed or going to roost, and a thin, high 'tswee'.

Robins are unusual in that during autumn and early winter both sexes defend territories. They are very pugnacious, and many territorial disputes take place, with much fluffing out of breasts to display the red patch to best advantage. About mid-winter the birds pair and thereafter share a territory. Garden robins are often quite tame, happily sitting close to the fork when digging is in progress, hoping for some titbit to be turned up. Away from gardens they prefer woods with a dense shrub layer – which explains why there are few in the Northern and Western Isles.

JUVENILE

37

Dunnock

Prunella modularis
Gael: *Gealbhonn nam Preas*

ADULT

Widespread   **All year**

This is the least conspicuous of the regular garden residents. *Mouselike in colouring*, it also behaves rather like a mouse, *creeping along* in an unobtrusive way. It does draw attention to itself, however, by frequent *wing flicking*, by its insistent, penetrating 'tseep' calls, and by its song, a thin warble, much shorter and less forceful than a wren's but quite musical. Like the wren it sings almost all year round and usually makes only short flights in the open.

Dunnocks are solitary and resident. They prefer a bushy habitat and frequent hedges and woods with a good layer of undergrowth, as well as gardens. They eat both insects and seeds, often foraging among fallen leaves, and feed more often under bird tables than on them. Apart from the Northern Isles and Outer Hebrides, where they are scarce or absent, they are widespread. Dunnocks are territorial like most small birds but research has shown that they do not follow the usual family pattern. Males outnumber females in spring and in some territories the female has two mates, both of which help to raise the chicks.

JUVENILE

ADULT

Wren

Troglodytes troglodytes
Gael: Dreathan-Donn

Widespread   **All year**

With its *pertly cocked tail* and unexpectedly *loud voice*, the *diminutive* wren is easy to recognise. Always on the move, it hops along the ground or climbs up stems, searching every nook and cranny for small insect life and frequently flicking its tail. Much of its time is spent in the shelter of bushes and low vegetation. Wrens sing almost throughout the year – a forceful and carrying, rather rattling, warble incorporating vigorous trills. They 'churrr' loudly when alarmed, and maintain contact with hard clicking 'tik-tik-tik' calls. They usually make only short flights before diving into cover.

Wrens are found in many different habitats: in gardens, woods and hedges, on rocky seashores and high in the hills. They seldom travel far, and as a result local 'races' have developed on some of the islands: Shetland wrens are bigger and darker in colour than mainland birds. The cock builds several domed nests but only the one chosen by the hen is lined. In severe winters many die, as insect food becomes hard to find; they are not attracted by bird table foods and do not move away to a warmer wintering area. In an effort to keep warm wrens sometimes roost communally in nest boxes or similar places – 43 were once counted coming out of a house martin nest!

TITS

Great (p.41), blue (p.42), coal (p.43), long-tailed (p.108), crested (p.109), marsh (p.110) & willow (p.110) tits

- are small, plump, agile and active, mostly hole-nesting, birds
- have short, fairly slim but strong bills, capable of hammering open seeds and nuts
- the sexes are similar and young are generally less brightly coloured than adults
- are all by nature woodland birds but the three commonest species (blue, great & coal) are now regular garden visitors/dwellers, while long-tailed tits also occasionally come to bird tables

The three scarcest species are regular only in different parts of the country:

marsh tit (p.110) in the southeast

willow tit (p.110) in the southwest

crested tit (p.109) in Speyside and around the Moray Firth

- marsh and willow tits might be confused with blackcap (p.101)

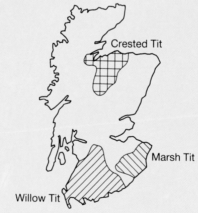

Great Tit

Parus major
Gael: Currac-Baintighearna

ADULT

Widespread   **All year**

This is the *largest of the tits* and one of the *most colourful* of garden visitors. Young birds have a brownish crown and a paler, rather washed out appearance. The great tit has a very varied repertoire of calls, most of which have a somewhat metallic sound; among the commoner are a ringing 'tink-tink', and 'tui-tui' repeated several times and rising towards the end. Its most easily recognised song is a loud 'teacher-teacher', metallic sounding like a saw being sharpened. Song starts in January and continues until about June.

Great tits frequent all types of woodland, scrub and hedges, but show a preference for deciduous trees; they are scarce or absent on many of the islands. They often feed on the ground, foraging among dead leaves for spiders, worms and insect larvae, and are adept at using their feet when feeding, not only to hold down large seeds such as beech mast and hazel nuts, but also to pull up and anchor strings holding peanuts. They are top of the tit pecking order, so the smaller blue and coal tits give way to them at the bird table.

JUVENILE

See 'Tits' p.40

41

Blue Tit

Parus caeruleus
Gael: Cailleachag a' Chinn Ghuirm

ADULT

Widespread   **All year**

This is one of the most frequent, and easily recognised, garden visitors. *Young blue tits, which are yellowish-green,* might be mistaken for warblers at first glance but have much shorter bills and a *darkish line through the eye and round to the nape.* The blue tit's song is a rapid, liquid trill, starting off with a few single 'tsee' notes. It has a variety of calls, though not as many as the great tit, the commonest being a slightly wheezy 'tsee-tsee-tsee-tsit'.

Blue tits are often very tame and will come to nut feeders fastened to a window pane, and remain sitting when a nest box is opened for examination. They inhabit woodland of all kinds, and are absent only from islands where such vegetation is scarce or lacking. They feed mainly on trees, from ground level up to the top, and only occasionally on the ground. In their natural woodland habitat the hatching of the young is synchronised with the main emergence of moth caterpillars, an ideal food for young chicks. In winter blue tits

JUVENILE   often flock with other tits in woodland.

See 'Tits' p.40

Coal Tit

Parus ater
Gael: Cailleachag a' Chinn-Duibh

ADULT

Widespread   **All year**

The coal tit, *smallest and dullest* in colour *of the three common tits*, has a *bold white patch at the nape* which distinguishes it from all other black-capped tits. Its range of calls is similar to those of great and blue tits but the notes are generally higher and thinner, while its 'see-chooo, see-chooo' song, with emphasis on the second note, is more piping than the great tit's 'teacher'.

Coal tits are low in the pecking order, giving way to both great and blue tits; they consequently tend to make only fleeting visits to bird tables, preferring to grab a peanut and take it away to eat in peace, rather than to feed on the spot and risk a confrontation. In their natural woodland habitat they sometimes store seeds behind tree bark. The coal tit is one of the few species which prefer conifer woodlands and has spread as a result of recent afforestation; it is now absent only from the Northern Isles and Outer Hebrides.

See 'Tits' p.40

JUVENILE

43

Swallow

Hirundo rustica
Gael: Gobhlan-Gaoithe

ADULT

Widespread   **April–September**

The swallow's *long tail streamers* distinguish it at once, whether in the air or at rest. Young birds are duller in colour and have shorter streamers. Swallows look short-legged and shuffle awkwardly when on the ground, but are swift and agile flyers; at the nest they often hover with tail fanned. They sing a lot, both in flight and when perched, and their pleasant twittering warble is one of the most welcome sounds of spring.

Most swallows frequent agricultural areas, where they nest on rafters in farm buildings and often return year after year to the same site. They feed mainly by catching insects in flight, usually over water or near animals, but sometimes pick them off roofs or plants. They start to arrive in early April, but in cold springs may be a week or two later, as their progress north is linked with the rising temperatures which bring about hatching of their insect prey. Swallows are much less sociable than house martins during the breeding season but before they set off for Africa in late summer large pre-migration flocks, sometimes numbering several thousand birds, gather in various parts of the country.

JUVENILE

ADULT

House Martin

Delichon urbica
Gael: *Gobhlan-Taighe*

Widespread   **Late April–September**

Its broad *white rump* and relatively *short, shallowly forked, tail* distinguish the house martin from the somewhat similar swallow. When on the ground, or clinging to a nest, the white feathering on the legs is conspicuous. House martins have a soft and sweet twittering song but do not sing nearly as much as swallows; their call is a clear but rather quiet 'chirripp'. When on the ground at mud-patches they waddle, often with wings and tail raised.

Like most species which feed upon flying insects, house martins do not reach Scotland until late April or even early May. They are sociable birds, feeding and gathering mud for nest-building in small parties, and often building nests close together under the eaves of the same house. Most house martins are associated with towns and villages, although small colonies also occur on more isolated buildings, and some nest on sea and inland cliffs. They are local or scarce over much of the highlands and islands. In late summer flocks gather in preparation for migration, and move south in waves, especially near the east coast, on their way to Africa for the winter.

ADULT

45

Swift

Apus apus
Gael: Gobhlan Mòr

ADULT

Widespread   **May–August**

The swift's long *narrow rather sickle-shaped wings*, shrill *screams* and fast, *dashing flight* make it an easy bird to identify. These are truly birds of the air; they feed and mate on the wing and land only at the nest site. In flight they hold their wings stiffly and beat them very rapidly. The harsh screaming calls are given mainly in flight, but also sometimes by birds on the nest.

Swifts are entirely dependent upon flying insects for food, and in consequence are among the latest of the summer visitors to arrive, most reaching Scotland about mid-May. From then until early August 'screaming parties', chasing wildly around above the roof-tops, are a familiar sight in many Scottish towns and villages, but as soon as the weather turns cool the swifts depart. They nest in buildings, usually under the eaves of older two-storey houses or in towers, but will also occupy specially designed nesting boxes. Swifts are commonest in lowland areas, rather local in the highlands, and absent from the islands. In warm spells they are sometimes seen high up in the mountains.

FARMLAND

Farms in the hills and glens often have rough and rushy fields suitable for nesting waders

Farmland is an important habitat for birds, providing a variety of food resources – seeds of different sizes, insects and worms, and growing greenstuff – which suit a wide range of species. All farmland is not the same, however, and its bird populations vary with the type of farming, the extent to which the fields are intermixed with woods and hedges, and the geographical location. Broadly speaking there are two main farming types: arable land, which is usually intensively cultivated, and grazing ground, much of which is permanent, often rough, grassland. Although they have some birds in common, these types also have their own distinct specialities. The croft ground in the northwest and on the islands is really a class by itself, with different bird populations from farmland in other parts of Scotland.

ARABLE LAND

Intensively cultivated arable land lies mainly in the eastern lowlands, along the Solway, and around the Clyde valley and Ayrshire coast, but crop growing is not, of course, restricted to these areas and most farms have some arable ground. Both the crops grown and the time of year influence the birds likely to be seen, as they determine the amount and kinds of feeding available.

Autumn is the season of plenty, with both spilt grain and weed seeds abundant on stubbles. Pheasants, partridges (grey and red-legged), pigeons (stock doves as well as woodpigeons), finches, buntings and grey geese are among the birds that frequent the stubble fields, and in some areas also whooper swans (p.123). Potato fields, once the foliage has died down, are used as resting

One of winter birdwatching's highlights is seeing the huge flocks of geese flighting between their roosts and the fields over which they feed

The farmland with trees and hedges on the right of this picture supports more, and a greater variety of, birds than the barer fields on the left

areas by flocks of lapwings and golden plovers (p.170), and waste potatoes are later gleaned by geese, mallards (p.126) and whooper swans. In severe winters anything green showing above the snow, such as kale, cabbages or turnips, is likely to attract large numbers of woodpigeons, and greylag geese also attack turnips. Nowadays ploughing starts early in the autumn and brings many black-headed (p.143) and common gulls (p.142) to search for worms along the newly turned furrows.

In early spring young cereals are grazed by geese and whooper swans, whose large feet may do a lot of damage by puddling and compacting the wet soil. Lapwings and sometimes oystercatchers nest on the almost bare ground between the rows of young corn – and are likely to lose their eggs when the crop is harrowed or rolled. What other birds are present in spring and summer depends very

49

much on whether or not arable fields have hedges round them, and patches of woodland, or even scattered trees, nearby. Without such nesting sites within easy reach the range of species is limited; with them there is likely to be a good variety of breeding birds, including some primarily woodland birds, such as finches, thrushes, robins (p.37), wrens (p.39) and dunnocks (p.38). Yellow-hammers, tree sparrows and whitethroats (p.99) are typical hedgerow birds, while corn buntings – now restricted to a few areas in the east – and grey partridges are the species most dependent upon the combination of hedges and arable land. Hedges and hedgerow trees are also important to wintering fieldfares and redwings, both as a source of berries for food and as a safe retreat when disturbed while feeding over fields.

Well-grown hedges containing a mixture of trees and shrubs provide good nesting sites as well as winter food supplies

GRASSLAND

From a nesting bird's point of view, rough or damp grassland, with tussocky growth to provide cover, is a better habitat than very short, heavily grazed, pasture. Rough grassland is common on farms in Scotland's glens, and also where lowland farms have poorly drained areas. Such land is important for breeding waders, such as lapwing, curlew and redshank (p.208), and other ground-nesting species like skylark (p.173) and meadow pipit (p.173). Snipe (p.146) are also likely to be present in the dampest areas, the usual range of hedgerow birds wherever there are hedges, and also in some areas magpies (p.24).

Where feeding is concerned, it is probably an advantage to many birds if the grass is short. Some of the species that feed over grassland do so by probing into the soil in search of insects, worms and other small creatures; the depth to which they can probe depends partly upon their bill shape and length, and partly upon how firm the soil is. Rooks, jackdaws (p.23) and starlings (p.26), which often feed together, cannot reach very far below the surface, whereas the longer-billed waders can probe quite deeply when the ground is damp and soft. Some species, including meadow pipits and skylarks, take insects on the vegetation; some, such as swallows (p.44) and the small gulls, hawk over the fields catching flying insects; and some eat the grass itself.

Many species can be seen on grassland under special conditions or in particular areas. Greylag, pink-footed, white-fronted and barnacle geese regularly graze over grassland, as do wigeon (p.199) along river banks. Where poorly drained areas

Wildfowl, waders and gulls are quick to gather on flooded fields in the winter

flood in winter, the standing water attracts gulls (p.188) and ducks (p.124-5) and sometimes herons (p.145), moorhens (p.138), whooper swans and waders (p.203), while untended fields with patches of seeding thistles and other weeds are likely places for finches, especially goldfinches. Two very local birds of grassland are the yellow wagtail, which breeds in hayfields in Ayrshire, and the chough, for which the heavily-grazed coastal grasslands of Islay are the principal feeding ground.

CROFT LAND

A characteristic of most crofting areas is the intricate mosaic of grassland, cropped ground and fallow land, all on a small scale and often closely associated with stretches of machair (see p.181), bog and short heather moorland. This mixture of vegetation, wetland and bare ground, together with the largely traditional farming practices still followed, provides suitable habitat for some species not usually found on farmland elsewhere in Scotland. The best-known of these is the corncrake, now confined as a regular breeder to some of the islands, where it makes its presence obvious by its monotonous night-long 'craking'. Many small waders breed on the machair, while in

Weedy areas in fields and on waste ground attract seed-eating species such as goldfinches

most crofting areas the commonest finch is the twite (p.176), a bird of hill country on the mainland but which often feeds over croft ground and nests on nearby heather moor.

OTHER FARMLAND BIRDS

Farm buildings, which still often include open sheds and barns, offer nesting sites for barn owls, which hunt over the fields, and birds such as swallows (p.44) and pied wagtails (p.36), which feed on the rich insect life often associated with livestock. Modern buildings are less good for birds than old stone built steadings, however, as they do not have the same abundance of rafters, ledges and holes. The small birds and mammals which feed on farmland regularly attract kestrels and in winter hen harriers (p.164), while sparrowhawks (p.86), buzzards (p.163) and the recently reintroduced red kite will hunt over open ground with scattered woodland. Crows often nest in solitary trees along a hedgerow or in a small wood, and scavenge whatever may be available in the way of food, which on hill farms may include dead sheep and also feedstuffs put out for livestock.

GEESE

White-fronted (p.55), pink-footed (p.56), greylag (p.57),
barnacle (p.58) & canada (p.124) geese
- are mainly winter visitors
- usually feed on fields by day and roost on water or
  moorland at night
- are very gregarious and feed in close flocks
- often fly in wavering lines or V formation
- always have some alert 'lookouts' when feeding
- differ in pattern &/or colour of head and neck, size &
  colour of bill, and colour of legs
- some have distinctive wing colours (visible in flight) or
  breast markings
- all show white on rump/tail in flight
- with experience, can often be identified by their calls

White-fronted Goose

Anser albifrons
Gael: Gèadh-bhlàr

ADULT

Local   October–early April
This species is distinguishable, at all ages, by its *orange legs*
and *smallish orange-yellow bill*. Young birds lack the adults'
white forehead and *black bars across the belly*. In flight head,
neck, back and wings look uniform in colour. White-
fronted geese winter in the west: Islay
holds the largest numbers, but there
are regular flocks on several other
islands, the Kintyre peninsula, and in
the south-west. They feed over rough,
boggy ground as well as farmland, and often
roost on moorland lochs. Flocks are usually
small, seldom more than 100 birds.

Pink-footed Goose

Anser brachyrhynchus
Gael: *Gèadh*

Widespread   **Late September–April**

Pink-footed geese can be distinguished from the larger and heavier greylag by the contrast between their markedly *dark brown head and neck and grey-brown body*, and their rather *small, mainly pink bills*. Their usual call, often heard when the birds are in flight, is a high, occasionally squeaky, two to three syllable 'pink-pink'.

Pink-footed geese often feed and roost in very large flocks, at times numbering several thousand birds, and are warier than greylags. They start to arrive here from Iceland during the second half of September, feed over rich farmland and roost on lochs and estuaries, or – less often – on moorland; most of their roosts are long-established. On arrival they glean grain, especially barley, from stubbles; later in autumn they move onto cleared potato fields and grassland; and towards spring they feed mainly on grass but sometimes on young cereal crops. They are most numerous in lowland east and central Scotland, from the Borders to Easter Ross, but some also winter around the Solway Firth.

See 'Geese' p.55

Greylag Goose

Anser anser
Gael: Gèadh-Glas

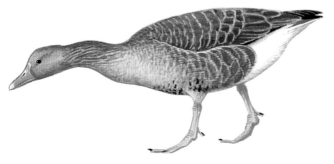

Widespread October–April   **Local in summer**

A *uniformly grey-brown* goose, with pink legs and a *heavy bright orange bill*, this is the ancestor of the farmyard goose. In flight it shows *silvery-grey 'shoulder' triangles* which contrast with the dark hind edge and tips of the wing. The greylag's voice is much deeper than a pink-foot's, a tri-syllabic 'ang-ung-ung', rather like a farmyard goose.

Greylags from Iceland arrive from early October onwards. They sometimes mix with pinkfeet when feeding but usually roost separately, tend to be in smaller flocks and use a wider selection of roosts, sometimes spending the night on a river bank. Immigrants winter in much the same areas as the pink-footed geese, but are more widely scattered. They occasionally damage turnip crops in hard winters, when other food is difficult to obtain. This species also breeds in Scotland, as a native resident in the Western Isles and northwest highlands, and as a feral population (descendants of birds released from captivity) scattered widely from Easter Ross to the Solway. These birds remain in or near their breeding area throughout the year.

See 'Geese' p.55

Barnacle Goose

Branta leucopsis
Gael: Cathan

Local   **October–April**

A *pure white face and forehead and jet black neck* at once distinguish the barnacle goose. Its back is grey barred with black, and its underparts are silvery grey, but the general effect is of a smart black and white bird. Its legs and small bill are black. Seen from below the black neck and upper breast contrasts sharply with the whitish belly. This species is very gregarious, feeding and flighting in tight packs, and is noisy and quarrelsome; at a distance a flock sounds like a pack of yapping lapdogs!

Wintering barnacle geese are less widely distributed than the grey geese, and occur mainly in two distinct areas. Those from Spitzbergen frequent the shores of the Solway while those breeding in Greenland visit islands off the north and west coasts, with by far the biggest concentration on Islay. Elsewhere in Scotland barnacles occur on passage or as stragglers among grey geese. The Solway birds start to arrive in late September and the Islay flock in the second half of October; both leave again in April. Barnacle geese feed on coastal saltmarsh (merse) as well as on farmland.

See 'Geese' p.55

Pheasant

Phasianus colchicus
Gael: An Easag

FEMALE

MALE

Widespread **All year**

Although really woodland birds, pheasants are often seen feeding in fields or walking along roadsides. Their *very long tails* make males easy to identify, but their colours vary as a result of inter-breeding with decorative species; very dark birds are not uncommon. After giving their strident crowing 'korrk-kok' call the males may whirr their wings briefly. Pheasants often seem reluctant to take wing and will run to and fro beside a fence rather than fly over it. When they do decide to fly the take-off is noisy, and they usually go only a short distance, travelling low and straight, before seeking cover.

Pheasants are gregarious; in winter the sexes sometimes form separate flocks. Many are artificially reared and released for shooting, and large numbers are often present in autumn close to rearing and feeding sites. They typically feed in a crouched position, frequently raising their heads to check for danger. This species, which was brought to Britain about 400 years ago, is still being introduced in some parts of Scotland, but it is essentially a lowland bird and is scarce in the central and northwest highlands and the islands.

Grey Partridge

Perdix perdix
Gael: Cearc-Thomain

Widespread   **All year**

Its *orange face and throat* immediately identify this *dumpy* bird, whose short neck and *drooping tail* give it a rather sad hunched appearance. In flight a dark inverted horse-shoe mark on the pale belly can be seen. Young birds lack colour on face and underparts, and are streaky brownish all over. The calls of the grey partridge are a high creaky 'kree-vit', and a rapid cackled 'ek-ek-ek' as it takes flight. Family parties stay together, as 'coveys', well into the winter, and although wary are easy to watch from a car; they feed most actively early and late in the day. When alarmed they crouch then run for cover; if forced into flight they usually stay close together and fly low.

Grey partridges feed over the fields and nest among rough growth along the bottom of hedges. Hedge removal has reduced potential nesting sites, and spraying of crops has resulted in such a decrease in insect life that chicks hatched on purely arable land have difficulty in finding enough food to sustain them. Because it is so dependent upon farmland, this species is absent from most of the highlands and islands.

Red-legged Partridge

Alectoris rufa
Gael: *Cearc-Thomain Dhearg-Chasach*

Local **All year**

This is a more *strikingly coloured and patterned* bird than the similarly shaped but slightly smaller grey partridge. Young birds have only indistinct face markings. The red-legged partridge's calls are a harsh 'shrek' and a more conversational 'chuk-chuka'. An introduced game-bird, this species is reared and released locally in various parts of lowland Scotland. It sometimes comes close to farms to feed with free-range hens. Not many survive our winters but a few occasionally do so and manage to breed in the wild for a few seasons.

Corncrake

Crex crex
Gael: *Trèan-ri-Trèan*

Local **Late April–early September**

Corncrakes are much more likely to be heard than seen, proclaiming their presence night and day with a monotonous and *distinctive rasping* 'crek-crek'. They are shy birds which skulk among long vegetation, in meadows, marshy areas and weedy patches of arable fields, and seldom come out into the open. Once widespread on farmland, this species now breeds regularly only in the western islands, mainly on the Outer Hebrides and Tiree, where traditional crofting methods still provide the right kind of habitat.

Curlew

Numenius arquata
Gael: Guilbneach

Widespread   **All year**

The curlew's *voice* and *long down-curved bill* make it easy to recognise. Its song starts with several slow deep whistling notes and then accelerates into a long liquid bubbling or rippling trill; this is one of the most familiar spring sounds in Scotland's glens, but can actually be heard almost throughout the year. The equally familiar ringing 'coor-lee' is the usual contact call, used on the breeding ground as the birds circle and glide high in the air, and also in wintering areas and while in flight on migration.

In the breeding season curlews frequent farmland and the lower hills, usually nesting on rough grazing or damp moorland but occasionally on arable fields. They appear on the breeding grounds from late February and start to leave again towards the end of June. By early August flocks have gathered at the coast, where they feed over the mudflats and roost on fields nearby. Very large numbers winter on farmland in Orkney, where the relatively frost-free climate allows them to feed without difficulty; elsewhere they are dependent upon being able to probe deep into estuarine mud for food.

See 'Waders' p.203

Lapwing

Vanellus vanellus
Gael: *Curracag*

MALE SUMMER

Widespread   **All year**

Lapwings are probably the most familiar of the waders, with their *wispy crests*, contrasty *black above and white below* plumage, and loud nasal 'pee-wit' or 'pees-weep' calls. In flight their *wings look very rounded*, and their wing beats are noticeably noisy. In the breeding season they perform aerobatic display flights, involving steep ascents, sudden dives, slow 'butterfly' flapping, and much rolling from side to side, when the birds look alternately light and dark.

Lapwings are more strictly birds of farmland than other waders, nesting on rough grassland or arable fields and wintering mainly on low-lying coastal grassland rather than on mudflats. When feeding they walk about, stopping frequently with head tilted forward, to pick up or probe for small invertebrates. They nest solitarily but start to flock from mid-June onwards, often with golden plovers (p.170), on stubble, short grassland and potato fields. As winter approaches Scottish birds move nearer the coast, or leave for areas where there is less likelihood of frost, while immigrants arrive from further north. By mid-March most of our native lapwings are back on their breeding grounds.

See 'Waders' p.203

Rook

Corvus frugilegus
Gael: Ròcais

ADULT

Widespread   **All year**

Rooks are *glossy black* all over, with *loose feathering on the
'thighs'* which gives them a 'baggy breeks' appearance.
Adults have dagger-shaped *greyish bills, at the base of which is
a bare whitish patch*; young birds have black bills and fully
feathered faces, making them look more like carrion crows.
Rooks are very gregarious and are nearly always in flocks,
whereas carrion crows are solitary. The rook's voice is less
harsh and croaking than a crow's; its calls include a varied
range of 'kaa' and 'kaw' notes.

This species is very closely associated with farmland and
is consequently absent from much of the highlands and
islands. Rooks feed over open ground, often in company
with jackdaws (p.23), and eat cereals, worms and insects.
Their dusk roosting flights are very obvious; the birds
assemble on the ground before flying, in a low and leisurely
straggling stream, to their woodland roosts.
Winter roosts may hold thousands
of birds, drawn from several
separate breeding rookeries in
the surrounding area. A small wood
or strip of trees seems usually to be
preferred for breeding; the untidy
stick nests are sited high in the trees.

ADULT

Carrion/ Hooded Crow

Corvus corone
Gael: Feannag

CARRION

Widespread   **All year**

Carrion crows are glossy black all over and can be distinguished from rooks by their heavier *black bills with no bare patch* and the absence of shaggy feathering around the upper legs. Hooded crows are pale grey on the back and belly. The crow's call is a deep hoarse 'kraak', usually repeated several times. Voice, and the *square ended tail*, are useful in distinguishing carrion crows from ravens (p .168) in the highlands and islands.

Carrion and hooded crows are more or less separated geographically: east of a line from Speymouth to the Inner Solway carrion predominate; west of a line from Bettyhill to Bute most are hooded; and in the north-south band in between both are present and inter-breeding occurs. Hybrids vary from dark grey on back and belly to almost black. Crows are catholic in their choice of habitat, occurring from towns to mountainous areas and the coast. They nest solitarily on trees and cliffs, and take a wide variety of foods. Much persecuted in the past, they are now much less hesitant about coming close to man, and regularly breed in towns and scavenge around rubbish tips.

HOODED

Chough

Pyrrhocorax pyrrhocorax
Gael: Cathag Dhearg-Chasach

Local   **All year**

Its *red legs and bill* distinguish this jackdaw-sized bird. Choughs are sociable and often indulge in aerial acrobatics. They sound rather like jackdaws (p.23) but their 'keeow' call is higher, more long-drawn-out and more musical than the jackdaw's 'kya'. Most of the small population of choughs is on Islay, around the Rhinns and the Oa, with a few on Jura, Colonsay and the Kintyre peninsula. This species nests in caves and derelict buildings, and feeds on invertebrates in the soil, especially on cattle-grazed grassland.

Red Kite

Milvus milvus
Gael: Clamhan Gobhlach

Local

All year

Their *deeply-forked tails* and narrow, strongly angled wings readily distinguish red kites from buzzards (p.163), which have a similar mewing call. They fly buoyantly and soar effortlessly, and are most likely to be seen in open country with scattered trees or small woods. In the past this species was exterminated by persecution but it has recently been re-introduced, with the release of young birds starting in the highlands in the late 1980s; some are now breeding. After the breeding season the birds scatter widely and may be seen almost anywhere in the country.

See 'Birds of prey' p.161

Kestrel

Falco tinnunculus
Gael: Clamhan Ruadh

ADULT MALE

Widespread   **All year**

The kestrel is the only small bird of prey which spends much of its time *hovering, with flickering wings and depressed tail*; on sighting prey it drops suddenly and quickly to seize it on the ground. When not hunting kestrels often *perch in an upright position* on telegraph poles and dead trees. Heavy barring with black makes the female's tail, back and wings look much darker than the male's. In direct flight the wing beats are fast and shallow, the tail looks long and narrow and the *wings long and pointed*. Kestrels frequently soar, with tail fanned. They are solitary and usually silent, but near the breeding site give a shrill high 'kee-kee-kee' call.

ADULT FEMALE

   The most abundant and widely distributed of the raptors, kestrels are residents wherever there are suitable sites for nesting, eg buildings or quarries with ledges, large tree holes, or old crow nests. They generally hunt over open ground, such as farmland and the verges of motorways, less often in open woodland and parks or over waste ground in towns. They are scarce only in the Northern Isles and Outer Hebrides.

See 'Birds of prey' p.161

67

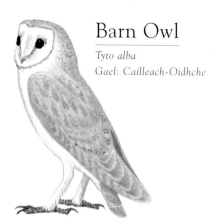

Barn Owl

Tyto alba
Gael: *Cailleach-Oidhche*

Local **All year**

Barn owls are most likely to be seen in the beam of head-lights as they hunt by night along a roadside, or roosting by day in a building. They are largely nocturnal but sometimes hunt in late afternoon. This is the *palest of the owls*, with a large round head, a *heart-shaped white face* and *dark eyes*. Its long, *white-feathered* and rather 'knock-kneed' *legs* dangle when it flies and are clearly visible as it stands upright on a perch. Its hunting flight is low, wavering, buoyant and absolutely silent. At the nest it hisses and 'snores'; its main call is an eerie shriek.

Barn owls are most abundant in the Borders and southwest, where they hunt for voles and mice in young forestry plantations, and over hill sheep ground as well as low-lying arable land. They nest in tree-holes or on a ledge inside a building; on some farms special shelves have been put up in modern barns to provide suitable sites. They are rather scarce throughout the central lowlands and up the east coast, and absent from most of the highlands and islands.

Woodpigeon

Columba palumbus
Gael: Calman Coille

Widespread   **All year**

This is the only pigeon with a *white patch on the side of the neck* (absent in young birds) and a *broad white crescent*, conspicuous in flight, from front to back *on the wing*. The woodpigeon's song is a lazy sounding 'coo cooo coo coo-coo', with the emphasis on the second note; the stock dove stresses the first syllable.

ADULT

Woodpigeons are widely distributed and abundant; they sometimes damage crops and can be a serious pest, especially where woods close to farmland attract large roosting flocks in winter. They are noisy birds, 'exploding' when disturbed with a loud clatter of wings; their display flight is also noisy, involving loud wing claps.

JUVENILE

Stock Dove

Columba oenas
Gael: Calman Gorm

Widespread   **All year**

Stock doves *lack the woodpigeon's white marks* and can be confused with feral pigeons (see p.211), which do not always have white rumps. They often feed with woodpigeons in winter but are less numerous and likely to be seen only in small groups. Stock doves are found on low ground from the central lowlands southwards and up the east coast to the Moray Firth. They feed among trees and shrubs as well as in the fields, but do not occur in town parks or gardens.

ADULT

MALE

Yellow Wagtail

Motacilla flava
Gael: Breacan Buidhe

Local   **May–September**

Yellow wagtails have *green, rather than grey, backs and no throat markings*, unlike the longer-tailed grey wagtail (p.150). The yellow wagtail's habitat is also quite different; it favours damp meadows, marshy ground, hayfields and occasionally cornfields, whereas the grey wagtail is very much a bird of fast and rocky streams and rivers. Yellow wagtails nest regularly only in a few places in low-lying parts of Ayrshire, Lanarkshire, and the Borders.

THRUSHES

Blackbird (p.27), song thrush (p.28), mistle thrush (p.94), fieldfare (p.71), redwing (p.71) & ring ouzel (p.172)

- are roughly blackbird size
- feed on the ground, on worms & insects, and on berry-bearing trees and shrubs
- are mainly residents or winter visitors (ring ouzel is a summer visitor)
- most are gregarious outside the breeding season
- sexes are alike or nearly so
- habitat, season, behaviour, general colouration and the presence of distinctive coloured areas are all helpful in identification

Fieldfare

Turdus pilaris
Gael: Liath-Truisg

Widespread   October–April

The *grey head and rump* immediately distinguish this rather
rakish-looking winter visitor. Fieldfares are noisy, giving their
harsh chattering 'chak-chak-chak' calls often both in flight
and at rest. They look alert and hold themselves rather upright
when perched and on the ground. They spend a greater
proportion of their time feeding in the open on fields
than do the smaller redwings, with which they often
flock, but also take berries from the hedgerows. In
hard weather fieldfares sometimes come into
gardens, and readily take fallen apples.

Redwing

Turdus iliacus
Gael: Sgiath-Dheargan

Widespread   October–March

Redwings can be distinguished from
song thrushes (p.28) by their *cream
eye stripe* and the *chestnut on flank and underwing*, which is
very obvious in flight. Their call, a high thin 'see-ip', can
often be heard on foggy nights in autumn as flocks pass
overhead. This species is predominantly a winter visitor,
though a few pairs breed in this country; many move on to
southern Europe before severe weather sets in. Redwings
roam over farmland and hedgerows, often in
company with fieldfares, and in hard winters come
into gardens for cotoneaster and other fruits.

See 'Thrushes' p.70

Tree Sparrow

Passer montanus
Gael: Gealbhonn nan Craobh

ADULT

Widespread **All year**

Tree sparrows differ from house sparrows (p.30) in having a *chestnut crown* and a *black spot on the cheek* and in the sexes being alike. They are also smaller and slimmer, and have smaller and tidier bibs. Young birds have smudgy marks in place of the black spot and bib. The tree sparrow's flight call, a rather hard and hoarse 'tek-tek', is distinctive, as is one of their other calls, a short metallic 'chik'.

This is a bird of low cultivated ground with hedges or scattered trees. It feeds on grain and seeds and is sociable, often occurring in small flocks and occasionally joining up with finches and house sparrows in winter. Generally a hole-nester, it readily occupies nest boxes and sometimes breeds semi-colonially, with several pairs building close together in crevices of farm buildings; where holes are scarce it will build a dome-shaped nest in a hedge or among ivy.

Tree sparrows are most abundant in the Lothians and Fife, scarcer but widely distributed over the rest of lowland Scotland, and absent from the islands and mountainous areas.

JUVENILE

See 'Sparrows' p.29

Goldfinch

Carduelis carduelis
Gael: Deargan Fraoich

ADULT

Widespread   **All year**

Adult goldfinches of both sexes are unmistakable with their *bright red faces and black and white heads*, while the flight pattern of conspicuous *yellow wing bars and a white rump* distinguishes this species at all ages. Young birds are streaky grey-brown, with no distinctive head pattern, and when at rest can be confused with several other species (see p.29). Goldfinches have a flitting, dancing flight and a liquid, trickling 'tswitt-witt-witt' call, which is incorporated in the canary-like twittering song.

Feeding largely on weed-seeds, this species is found wherever there are patches of weedy ground, on neglected farmland, along roadsides and in gardens. Goldfinches are remarkably agile little birds, able to hang almost tit-like at thistle and ragwort heads, and to pull in a stem with their bills and then use their feet to anchor it as they feed. They are not particularly shy and readily perch on wires etc. in the open. They are commonest, though nowhere very numerous, from the central lowlands southwards and absent from the islands and most of the highlands. Goldfinches are sociable but flocks seldom exceed 100 birds.

JUVENILE

See 'Finches' p.29

73

Yellowhammer

Emberiza citrinella
Gael: Buidheag Luachrach

MALE SUMMER

Widespread   **All year**

Yellowhammers are buntings with stout bills. Adults have *yellow underparts* and *streaked, chestnut brown backs*; in males the head and breast are bright yellow, while in females they are duller and more heavily marked with dark brown. Young birds are streaky yellowish brown. In flight the combination of *chestnut rump* and *conspicuous white outer tail feathers* is distinctive at all ages. The yellowhammer's song is often rendered as 'little bit of bread and no cheese', the emphasis usually coming on the 'cheese'; its calls are a brisk 'tink' and a liquid 'twitup', often given in flight.

Yellowhammers require prominent song posts, and make use of trees, hedgetops, fences and telegraph wires. They are birds of farmland with hedges, or other open ground with bushes and small trees, and occasionally nest along the edge of young conifer plantations. They feed mainly on the ground, taking seeds and grain. They are sociable after the breeding season and often flock with finches and sparrows in winter. Yellowhammers are common in most agricultural districts but are scarce or absent over much of the highlands.

See 'Buntings' p.29

FEMALE

Brambling

Fringilla montifringilla
Gael: *Bricein Caorainn*

MALE

Widespread   **October–March**

Bramblings look *like pale chaffinches* (p.31) at first glance, but can be distinguished by their *white rumps* and much paler bellies. The male's head and back are brownish in winter, becoming black by spring. Females have dark markings on their grey-brown heads. Bramblings have a distinctive hoarse, metallic 'tsweek' call; in flight they give a repeated 'chuk-chuk-chuk'.

Variable numbers of bramblings arrive from Scandinavia to winter in this country; in some years flocks several hundred strong are quite common, while in others very few appear. They are most likely to be seen on the eastern side of the country. Bramblings often flock with other finches, feeding over farmland and in woodland, especially on beechmast. In severe weather they sometimes come into gardens for seed but seldom land on bird tables.

See 'Finches' p.29

FEMALE

75

Corn Bunting

Miliaria calandra
Gael: Gealag Bhuachair

Local   **All year**

The corn bunting's *dumpy shape* and *short thick bill* are helpful in distinguishing it from the similarly *streaky brown* and much more widespread skylark and meadow pipit (p.173). Corn buntings often perch on fence or telegraph wires, or on posts, along field boundaries, and have a peculiarly *metallic jingling song*, like the rattling of a bunch of keys.

This species was at one time much more abundant on arable farmland but is now regular only along the coastal strip from East Lothian to Buchan, and on the islands of Tiree and the Uists, though a few still breed in other areas. It is believed that changing agricultural practices, especially the spraying of cereal crops, has made farmland a less suitable habitat for these birds.

See 'Buntings' p.29

WOODLAND

Highland birchwoods are the summer home of willow warbler, redstart, tree pipit and spotted flycatcher

Woodland is the natural home of many birds which now regularly inhabit gardens or farmland too, and also of others which have not been able to adapt their lifestyle in a similar way and cannot survive without the 'right' kind of woods. Just what is 'right' varies for different birds: for example those which excavate their nesting holes need dead trees, those which nest on or near the ground need the cover of shrubs or rough vegetation, and those which have very specialist food requirements need particular types of tree.

Natural woodlands, few of which now survive in Scotland, generally contain trees and shrubs of several kinds and varying ages, which provide both a varied food resource – nuts, cones, berries, insects, etc. – and shelter at different heights and densities. Many of today's amenity lowland woods, although they have a different range of trees and shrubs, offer similar diversity and support the most varied populations of woodland birds, and especially of summer visiting species. At the other end of the scale, dense mature stands of commercial conifers are of much less bird interest than the old native pinewoods, with their scattered

Estate and park woodlands planted for their amenity value often have a dense understorey of shrubs, which attracts some of the warblers

birch, rowan and holly. Clearings and rides within a wood make it more attractive to some species, as they give space for song flights and clear paths for hunting along. The character of a woodland consequently determines the birds likely to be encountered within it, and so too does its geographical location. For example, most warblers and flycatchers, typical summer visiting insect-eaters, become scarcer the further north you go, while capercaillie and crested tit are strictly birds of the highlands.

Watching woodland birds is seldom easy, as many of them spend much of their time hidden amongst the vegetation or high in the treetops, and it is here that recognition of songs and calls is especially useful. Early morning is the best time to hear bird song; some species become active earlier than others, so the dawn chorus involves a gradually increasing selection of songs.

LOWLAND BROADLEAF AND MIXED WOODLANDS

The more varied the composition of a lowland wood the greater the variety of birds likely to live in it, but the presence of wood warbler and pied flycatcher is particularly associated with oaks, which support more insects than any other tree. Dense shrubby cover, such as rhododendron or bramble, is important for garden warblers and blackcaps, thickets close to the wood's edge for whitethroats, and a combination of tall trees and some rough ground vegetation for chiffchaffs. Many estate woodlands have breeding woodcock, and a few in the southeast have marsh tits. Most woods have natural holes of varying sizes to serve as suitable nesting sites for tits, starlings (p.26) and tawny owls, but green and great spotted woodpeckers – and the very local willow tit – will only be present where dead trunks are available.

In winter, when many birds which feed primarily on insects in summer turn to seeds and fruits, beech trees become more important, with tits and finches gathering to feed on the crop of mast. Jays too take beechnuts, as well as acorns and hazelnuts.

BIRCH AND OTHER BROADLEAF WOODLANDS IN THE HIGHLANDS

Rather open birchwood, often containing few other tree species and little in the way of a shrub layer, is widespread over much of the highlands, sometimes straggling far up the hillsides. Two birds especially typical of such woodlands are the willow warbler, whose song seems to tinkle out from

Pied flycatchers and wood warblers are particularly associated with oakwoods which have little ground cover

Native pinewoods, with a mix of old and young trees, and heathery ground cover, are the haunt of crested tit, crossbill and capercaillie

amongst the leaves of almost every birch in summer, and the tree pipit, which gives voice as it parachutes down into a clearing from a treetop. Spotted flycatchers, redstarts and song thrushes (p.28) are also frequent inhabitants of these woods, along with the ubiquitous chaffinch (p.31), robin (p.37) and wren (p.39) and the three common tits, with long-tailed tits and redpolls also sometimes present, while sparrowhawks and cuckoos (p.172) move between the woodland cover and adjoining open ground. Even quite small clumps of broadleaf trees in the highlands are likely to hold a pair of mistle thrushes, and probably crows; slightly larger areas of woodland often have breeding buzzards (p.163). Many birds leave the higher highland woods in winter, at which season the birches and alders fringing river banks attract redpolls and siskins.

NATIVE PINEWOODS

This is the woodland type which long ago covered much of the highlands but now survives in only a few areas, most notably Deeside and Speyside, with smaller patches in Perthshire, Wester Ross and some of the glens to the west of Inverness. Scots pine is the dominant tree, but these woods also contain a generous scattering of birch, rowan, holly, gean and juniper, and have a tussocky ground cover of heather and blaeberry. Two specially Scottish birds are closely associated with this type of woodland, the very local crested tit, and the more widespread crossbill, the Scottish race of which occurs only in native pinewoods. The capercaillie, also originally a bird of these natural forests, has been able to adapt to life in other conifer woodlands and so extend its range, but is still largely dependent upon pine trees. Coal tits, goldcrests and redstarts are regular pinewood inhabitants, with tree pipits and spotted flycatchers where there are clearings with birches, while ospreys (p.141) often, and golden eagles (p.162) occasionally, choose to nest in Scots pine trees.

CONIFER PLANTATIONS

Today non-native conifers make up a very large proportion of Scotland's woodland, covering vast areas which were once moorland with forests of only one or two tree species. The bird populations of these woods change with their stage of growth. In the early stages, when there is still plenty of rough ground vegetation between the trees, some birds of open ground, such as skylarks (p.173) and meadow pipits (p.173), stay on and these are gradually joined by common woodland species such as robin (p.37) and chaffinch (p.31). Young plantations also provide suitable breeding ground for short-eared owls (p.169), hen harriers (p.164) and occasionally grasshopper warblers. By the time the trees are almost touching birds which like scrub have usually moved in, among them whinchats (p.175), yellowhammers (p.74), linnets (p.176), willow warblers and dunnocks (p.38), while most of the open ground species have left. At this stage whitethroats may be present at lower altitudes, and occasionally bullfinches (p.33) even on relatively high ground. As the woodland canopy closes over, the ground vegetation dies out and feeding opportunities for birds decrease, until the main food supplies available are tiny insects, which support goldcrests, treecreepers and coal tits (p.43),

Young conifer plantations provide suitable habitat for open-ground birds like meadow pipits, as well as small woodland species

Although few birds live in the heart of a well-grown plantation, a good many feed along its rides and fringes

and the shoots and seeds of the trees themselves, which are taken by black grouse (p.166), and by siskins and crossbills respectively. Clearings and rides allow sparrowhawks and other woodland species to continue to inhabit the woodland fringes, but the dense heart of a plantation supports few permanent inhabitants. It does, however, provide safe nesting and roosting places for birds such as woodpigeons (p.69), herons (p.145) and crows (p.65), which feed elsewhere in the surrounding countryside. Shelter belts and small conifer woods on moorland may be occupied by long-eared owls, and in the southwest nightjars sometimes summer in areas where conifers have been felled.

Capercaillie

Tetrao urogallus
Gael: Capall-Coille

MALE

Widespread   **All year**

Turkey-sized capercaillies are most likely to be seen in flight, as they crash away through the trees after being flushed; much of the noise is from their wing beats – they are surprising agile when flying. The huge dark males are unmistakable, but females might be confused with female red or black grouse (p.166-7); size and the capercaillie's reddish breast and *broad fan tail* are the main distinguishing features. The male's 'song' is a strange series of accelerating clicks, ending with a cork-pulling 'pop'; like black grouse they indulge in communal display. Some males are very aggressive during the breeding season and will attack people or vehicles.

Capercaillies breed in conifer woods, most often of mature pine, but sometimes feed on adjoining fields or among deciduous trees. They are found mainly in the eastern highlands, from Tayside to Easter Ross, with smaller numbers west as far as the Loch Lomond islands. Even without seeing them it is possible to know that capers are present by finding their characteristic droppings along a woodland ride; these are composed largely of conifer needles.

FEMALE

See 'Grouse' p.166

85

Sparrowhawk

Accipiter nisus
Gael: Speireag

ADULT MALE

Widespread   **All year**

Sparrowhawks have *short rounded wings, long barred tails, heavily barred underparts,* and *long yellow legs*; in flight they can sometimes be confused with cuckoos (p.172). The female is much larger than the male, brown rather than grey on the back, and whitish not red-brown below. They *hunt by flying low and fast* along hedges and among trees, and grabbing unsuspecting small birds with their feet. They often visit suburban gardens, creating panic among the usual residents, and may perch on fences or rooftops.

This species occurs on the mainland, and some of the inner islands, wherever there are woodlands thick enough to provide good cover but open enough to allow easy flight between the trees or along rides. In spring sparrowhawks soar high above tree level and circle in display. They are usually silent but when intruders are near a nest they give loud, rapid 'kek-kek-kek' calls. In the past they suffered from persecution and pesticide poisoning, but the population is probably now larger than ever before.

ADULT FEMALE

See 'Birds of prey' p.161

Woodcock

Scolopax rusticola
Gael: Coileach Coille

Widespread   **All year**

This *long-billed* wader is most likely to be seen in fairly open deciduous woodland or along the rides in conifer forests, as it performs its 'roding' display flight at dawn and dusk in spring and early summer. This takes it low over the trees, with slow almost owl-like wing beats; as it flies it gives a series of deep croaks followed by a shrill sneeze. Woodcock feed and nest on the ground in damp woodland with good ground cover; they are wary and seldom seen feeding, but sitting birds sometimes allow a close approach. They are beautifully camouflaged, their *heavily marbled red-brown plumage* blending with the dead leaves and bracken around them. Their *eyes are set high on the sides of the head*, so that they are able to watch for predators overhead. When flushed they fly away noisily, quickly dropping back into cover.

Woodcock are widespread on the mainland but do not breed on the Northern Isles or Outer Hebrides. Some move south or cross to Ireland in winter but many remain in Scottish woodlands all year.

See 'Waders' p.203

Tawny Owl

Strix aluco
Gael: Comhachag Dhonn

Widespread   **All year**

This is the commonest owl and is responsible for the familiar 'too-whit' and 'hoo-hoo-hoo' calls. It defends its territory throughout the year and regularly calls in October and November. Although primarily a woodland resident it also frequents parks and suburban gardens. Tawny owls are more often heard than seen, as they normally hunt between dusk and dawn, dropping silently onto their small rodent prey. Pellets of indigestible skin and bones on the ground beneath a tree show that owls have been roosting there.

Tawny owls are *tubby*, with large round heads, and are a rather *uniform tawny or greyish brown*, with paler facial 'discs'. Roosting birds can be distinguished from long-eared owls (p.89), the other species most likely to be found in a tree, by their *black eyes*, which give them a gentle expression, and the absence of ear tufts. Widespread on the mainland, they are most abundant in mature deciduous or mixed woods, but also occur among scattered trees and in conifer woodland; where nesting holes are scarce they readily occupy boxes.

Long-eared Owl

Asio otus
Gael: Comhachag Adhairceach

Widespread   **All year**

A mainly nocturnal hunter, this owl is more likely to be found at its roost than seen in flight. When disturbed at the roost it draws itself up like a cat, making its body appear *taller and slimmer*, while its *glowing orange eyes* and conspicuous *ear-tufts* distinguish it from the tawny owl (p.88). The long-eared owl's cry is a soft, moaning 'oo-oo-oo'; it is a much more silent bird, even during the breeding season, than the tawny owl. Young birds, when nearly or newly fledged, give loud squeaky calls which sound rather like a creaking gate; these are often the first sign of this species' presence.

Long-eared owls typically occupy small patches of conifer woodland surrounded by open hunting habitat, often moorland. Although widespread on the mainland, they are nowhere very numerous and are absent as breeding birds from much of the northern and western highlands and most islands. They usually lay in the old nests of other species, such as crows.

Great Spotted Woodpecker

Dendrocopos major
Gael: Snagan Daraich

MALE

Widespread   **All year**

The great spotted woodpecker's loud territorial *drumming*, produced by rapidly pecking at a branch, is often the first sign of its presence, and immediately identifies it. Its boldly *pied plumage* is also distinctive, and in flight, which is markedly undulating, its *white shoulder patches and red under tail* are conspicuous. Females lack the male's red nape patch, and young birds have red crowns. A sharp 'chik' call is given often, both while clinging to a tree and in flight. This species uses its tail as a prop as it chips away at dead wood to get at insects; scatterings of fresh wood chips show where a woodpecker has recently been at work.

Great spotted woodpeckers are widespread on the mainland and also occur on a few of the larger, well wooded islands. They frequent both deciduous and coniferous woodland, but can only breed where there are dead trees in which they can excavate their nest holes. In hard weather they sometimes visit bird tables for fat.

JUVENILE

Green Woodpecker

Picus viridis
Gael: Lasair Choille

MALE

Widespread   **All year**

This is the only *medium-sized* bird which is *predominantly green on back and wings*. Its yellow-green rump is conspicuous in flight and its *crimson crown* is also obvious. Green woodpeckers seldom drum; instead they make their presence known by their loud, carrying 'yaffle' call, which sounds like maniacal laughter and gradually accelerates as it descends the scale. They feed mainly on the ground and are especially fond of ants. Like the great spotted woodpecker, this species flies in deep undulating loops, closing its wings for several seconds between upward bounds. On the ground it holds itself rather upright and moves in hops.

A bird of low-ground deciduous woodlands, the green woodpecker requires a mix of mature trees and open ground. It started to breed in Scotland only in the 1940s and has since gradually extended its range; it now occurs as far north as Ross-shire and has nested on Islay.

JUVENILE

Jay

Garrulus glandarius
Gael: *Sgraicheag*

ADULT

Widespread   **All year**

Jays glimpsed in flight at the edge of a wood are easily identified by their *conspicuous white rump*, contrasting with *black tail*, *pinkish back* and rounded *dark wings* with a *vivid blue shoulder patch*. Jackdaw-sized, they look predominantly pinkish brown at rest, with white and blue patches on their dark wings; their black and white crown feathers are often raised to form a *crest*. Their usual call is a loud and very harsh 'kraak', given frequently as small parties move about in the woods.

Jays take a varied diet, including insects, acorns (which they sometimes store), eggs and young birds. They feed both on the ground and in trees, hopping about with jerking tails. Although found mainly in deciduous woods they may also be seen along the fringes of conifer plantations. Jays are most abundant in the southwest, Borders and Argyll, absent from much of the Clyde-Forth valley but abundant in the northern half of the central lowlands, scarcer in Fife and up the east coast to the Moray Firth, and absent from the highlands and islands.

ADULT

Nightjar

Caprimulgus europaeus
Gael: *Sgraicheag Oidhche*

Local   **Late May–early August**

This nocturnal, moth-eating summer visitor is more likely to be heard than seen. It *sings at night, with a loud fast churring*, which rises and falls and is often sustained. During the day it sits motionless on a branch or the ground, where its *cryptic colouring provides perfect camouflage*. Nightjars occur regularly only in Dumfries and Galloway and on Arran. They show a preference for rather open woodland, either deciduous or coniferous, with rhododendron or bracken and other ground cover in glades and rides. This species was more widespread in the past; its decline is believed to be partly due to climatic change, with late cold springs adversely affecting the availability of moths.

Mistle Thrush

Turdus viscivorus
Gael: Smeòrach Mhòr

ADULT

Widespread   **All year**

The mistle thrush is a *greyer brown* than the smaller song thrush (p.28) and has more *distinct large spots on its whitish breast*; in flight the *white 'corners' to its tail* show well. The absence of grey on head and rump distinguishes it from the fieldfare (p.71), which is similar in size. It looks bold and alert and stands rather upright. It sings from treetops, loudly repeating a number of short phrases; the fact that it sings in all weathers has earned it the name 'storm cock'.

This species is at home in any habitat where there are both trees and good expanses of open ground. It breeds in parks and large gardens as well as on farmland with small woods or scattered trees and on low moorland, but numbers are small and pairs usually well spaced out. After the breeding season mistle thrushes join up in small flocks; some leave the more exposed upland breeding areas in winter but most remain near their home ground. None breed in the Northern Isles and they have done so only occasionally in the Outer Hebrides.

See 'Thrushes' p.70

Tree Pipit

Anthus trivialis
Gael: Riabhag Choille

Widespread   **April–August**

Differences in voice and *choice of habitat* help in distinguishing this species from the similar meadow pipit (p.173). The tree pipit has a *loud and musical song*, more varied than the meadow pipit's, incorporating long trills and ending with a slow 'chew-chew-chew' or 'seea-seea'. It sings from the top of a tree or, more often, in *song flight*, which involves flying upwards from a perch, circling and parachuting back down. Its plumage pattern is like the meadow pipit's but it is *brown* rather than olive *above* and *yellower on the breast* which is streaked less heavily than a meadow pipit's. Young birds are more buff-brown on the back than adults. Tree pipits are found in open woodland or among scattered birch, pine or mixed deciduous trees, whereas meadow pipits are birds of open ground. Clearings within the wood, and tall trees from which to launch into song flight, are essential aspects of a tree pipit's habitat.

Although widely distributed over much of the mainland, this species is scarce in the arable farming areas of the central lowlands and east coast. It does not breed in the Northern Isles or Outer Hebrides. No tree pipits winter in Scotland.

Redstart

Phoenicurus phoenicurus
Gael: Eàrr Dhearg

MALE

Widespread   **May–early September**

The redstart is easily recognised by its *rusty red rump and tail, which it constantly quivers up and down* while on the ground and flirts fanwise in flight. The sexes are markedly different. Females are greyish brown above and pale orange buff below, and juveniles are similar but speckled like young robins (p.37). Redstart song is a rather robin-like short warble, but weaker and ending in a feeble twitter. The commonest calls are 'hwee-tuk-tuk' and a loud 'hweet', not unlike that of a willow warbler.

This species is found in relatively open woodlands, especially those including oaks and pines, and is most abundant in the central, west and northwest highlands, mainly in wooded glens. It breeds only locally in other parts of the mainland and not at all on most of the islands. Redstarts are hole nesters, but readily make use of nest boxes. They feed both on the ground and in the trees, occasionally hovering or flycatching in flight.

FEMALE

Warblers
- are roughly tit-sized but with proportionately longer tails
- are very active insect eaters, with slender bills
- are all migratory, arriving April/May and mostly gone by September
- voice & habitat preference are important in identification

Usually in trees: willow warbler (p.104), chiffchaff (p.103) & wood warbler (p.102); the 'leaf warblers'
- greenish with eyestripe, sexes alike
- song is the most reliable means of identification
- can sometimes be confused with goldcrest (p.105)

Usually in scrub/thickets: blackcap (p.101), whitethroat (p.99) & garden warbler (p.100)
- sexes alike in garden warbler, nearly so in whitethroat, quite distinct in blackcap
- whitethroat colouring & song distinctive; blackcap & garden warbler songs alike but colours very different
- male blackcap can be confused with marsh/willow tit (p.110)

Usually in or near marshy ground: sedge warbler (p.152) & grasshopper warbler (p.98)
- streaky brown backs, sexes alike
- song is the most reliable means of identification

Grasshopper Warbl

Locustella naevia
Gael: *Ceileiriche Fionnan-Feòir*

Widespread   **May–end July**

This *streaky-brown* warbler is much more likely to be heard
than seen. Its *song is a high-pitched mechanical churring* rather
like the winding of a fishing reel, frequently sustained for
1–2 minutes and sounding ventriloquial as the bird turns its
head. Grasshopper warblers sing both day and night, and
spend most of their time skulking amongst low vegetation
in young conifer plantations, reedbeds, or rough marshy
ground. The somewhat similar sedge warbler (p.152) has a
quite different song and is much more likely to be seen
perched in bushes or on reeds. Both have creamy eye-
stripes, but the grasshopper warbler's is faint whereas the
sedge warbler's is very conspicuous. They also differ in tail
shape, the grasshopper warbler's being rounded and the
sedge warbler's graduated to a point.

Although they may be present until well into August,
grasshopper warblers effectively 'vanish' when they stop
singing in late July. They breed regularly from the central
lowlands southwards, are scarcer elsewhere on the
mainland, and nest at least occasionally on some of the
inner islands.

See 'Warblers' p.97

Whitethroat

Sylvia communis
Gael: Gealan Coille

MALE

Widespread   **May–late August**

The whitethroat's *brilliant white throat, rusty wings* and *white outer tail feathers* immediately distinguish it from the other common warblers. The male's head and nape are grey, and the female's brown. Whitethroats are perky birds, which often cock their tails and raise their crown feathers to form a slight crest. They have rather harsh voices, with a variety of scolding 'charr' and repeated 'chak' notes, and the male's song, often given in a brief dancing flight, is a vigorous and somewhat scratchy chatter.

The whitethroat's habitat typically includes patches of tangled vegetation, brambles and small bushes set in fairly open ground, and it is often found around overgrown hedgerows and along woodland edges. This species can be difficult to spot as it spends much of its time in cover, usually searching for insects among foliage rather than on the ground, and makes only short flights in the open. Its voice is often the first indication of its presence. Whitethroats are most abundant in the central lowlands and southern counties, becoming scarce further north. They are among the later-arriving summer visitors, with many not appearing until well into May.

See 'Warblers' p.97

Garden Warbler

Sylvia borin
Gael: *Ceileiriche Gàrraidh*

Widespread   **May–August**

This is a shy and solitary bird, which skulks in cover and seldom flies in the open long enough to give a good view. Its *song*, delivered from cover, is consequently *important for identification* – a sweet, mellow and flowing, rather subdued but well-sustained warble, sometimes lasting for several minutes with only brief pauses. The blackcap's song is somewhat similar but more varied and shorter. Both have 'chek', 'chak' and 'churr' calls. Garden warblers are *rather plump* and have *no distinctive markings*, but can be distinguished by a process of elimination! Unlike many other warblers, they seldom flick their wings and tails.

Garden warblers require dense shrubby cover in their breeding area, and frequent patches of rhododendron or bramble growing as an underlayer in fairly open woodland, overgrown hedges and occasionally young conifer plantations at the thicket stage. They are commonest from the central lowlands southwards, especially towards the west; occur occasionally up the east coast to the Moray Firth; and are absent from the islands and most of the highlands. This is one of the later arriving of the insect-eating summer visitors.

See 'Warblers' p.97

Blackcap

Sylvia atricapilla
Gael: *Ceann-Dubh*

MALE

Widespread   **April–August**

Their *distinctive coloured caps and slim grey-brown bodies* make blackcaps easy to identify, though males might be confused with the black-chinned and much plumper marsh and willow tits (p.110). Blackcaps spend a lot of their time in cover or high in the tree canopy, and their song is often the first indication of their presence. They have a loud and carrying warble, clearer and more varied than a garden warbler's, but usually less sustained and in shorter phrases. The usual contact call is a rather sharp 'chak' or 'churr'.

Blackcaps are birds of mature deciduous or mixed woodland with good shrub cover, a combination typically found in parks and estate policy grounds. They are most abundant in the central and southern lowlands, regular up the west coast as far as Ardnamurchan, and occur occasionally in the Great Glen and near the east coast up to the Dornoch Firth. Although essentially summer visitors, a few blackcaps are present most winters; these probably did not breed here but arrived across the North Sea in late autumn. Wintering birds readily come to bird tables, and act quite aggressively towards other small species such as blue tits.

See 'Warblers' p.97

FEMALE

Wood Warbler

Phylloscopus sibilatrix
Gael: Conan Coille

Widespread   **May–August**

The wood warbler is the *largest and most green and yellow of the three small 'leaf' warblers*, and has a *distinct eye-stripe*. It is unusual in having two quite different songs, both of which are distinctive enough to identify it. One is a sustained shivering trill, which starts slowly and gradually accelerates, and the other a deliberate piping 'pyu-pyu-pyu' repeated seven or more times. Both are loud and penetrating, carrying quite a distance and rapidly disclosing the presence of this species in a wood. Wood warblers spend most of their time high up in the tree canopy and often allow their wings to droop loosely as they move about.

This species prefers mature deciduous woods with little ground cover, and is especially associated with oakwoods. It is most abundant in west central Scotland, from Perthshire through Stirling to Argyll, and in Dumfries and Kirkcudbright. Elsewhere on the mainland it occurs only very locally or breeds only irregularly, and it is absent from most of the islands. Like other 'leaf' warblers, it builds a domed nest.

See 'Warblers' p.97

Chiffchaff

Phylloscopus collybita
Gael: *Caifean*

Widespread   **Late March–September**

This small warbler is *most surely identified by its song*, which is composed of two distinct notes, 'chiff' and a lower 'chaff', repeated monotonously but in variable sequence. In appearance it is very like the much commoner willow warbler (p.104), but slightly more brownish than greenish above and buff rather than yellowish below.

Chiffchaffs favour woods and parkland where tall trees, either deciduous or coniferous, are intermixed with shrubby growth; they spend much of their time high in the canopy. They are restless and active feeders, hopping and flitting along the branches with constantly flicking wings and tails, and occasionally hovering or making short flycatching flights. This species is scarcer and more local than the willow warbler, breeding mainly in the southern half of the country but at least occasionally almost anywhere that has suitable habitat. Chiffchaffs are the first of the warblers to arrive in spring. Quite large numbers of migrants sometimes reach the east coast late in autumn, after the Scottish breeding birds have left, and a few of these sometimes stay throughout the winter.

See 'Warblers' p.97

Willow Warbler

Phylloscopus trochilus
Gael: Crionag Ghiuthais

Widespread   **April–August**

This little 'leaf' warbler is responsible for much of the bird song to be heard in Scottish woodlands in summer. *Its song immediately confirms its identity and is easy to recognise*: a rather plaintive sequence of silvery liquid notes descending the scale, usually rather faint at the start and becoming more deliberate. Its contact call is a soft 'hooeet'.

Willow warblers are found in all types of woodland, other than well-grown conifer plantations, but are especially typical of birchwoods, and are often present even where there are only a few scattered birches on a hillside. They breed throughout the mainland and on many of the islands, but not regularly in the Northern Isles. They feed more often at low level than do chiffchaffs (p.103). During the autumn migration, from July onwards, waves of willow warblers drift southwards, often turning up in gardens and town parks where they do not breed. These passage birds sometimes sing – a rather muted version of the full song – and at this season young birds tend to look much brighter in colour, with decidedly yellow underparts.

See 'Warblers' p.97

Goldcrest

Regulus regulus
Gael: Crionag Bhuidhe

ADULT

Widespread   **All year**

The goldcrest is the *smallest British bird*, and is dumpy-looking with *no obvious neck, a tiny bill, very large eyes for its size* and *a distinct wing bar*. Young birds lack the golden 'crest' and might be mistaken for a willow warbler (p.104), which is longer and slimmer, and has no wing bar. Goldcrests spend much of their time high up in the trees, where their thin, shrill 'zee-zee' calls disclose their presence. Their song is also very 'thin' and squeaky, a repeated 'seeter-seeter' ending in a flourish; it is so high-pitched that some people cannot hear it.

Goldcrests are found in most types of woodland, including even small shelter belts, but are most abundant in conifer woods. In winter they often join flocks of tits and treecreepers, roaming through the trees in search of minute insects. A combination of snow and frost is likely to result in many deaths, as these tiny birds are unable to find sufficient food during the limited daylight hours to keep them alive through a freezing night. Scottish breeding goldcrests are resident, but large numbers of immigrant birds sometimes arrive on the east coast in autumn.

Spotted Flycatcher

Muscicapa striata
Gael: Breacan Glas

Widespread **May–early September**

Flycatchers characteristically sit very upright, on a prominent perch, from which they make short flights after insects, often returning at once to the same perch. They have flattish, broad-based bills, large dark eyes and rather short legs. The *rather nondescript* spotted flycatcher is the larger and by far the commoner of the two species which breed in Scotland. The sexes are alike; young birds have a scaly appearance to head, back and breast. The spotted flycatcher's voice is thin and squeaky: its call is a scratchy 'tzee' and its insignificant song comprises only five or six notes. Its flight is fluttering and erratic, with rapid twists and turns.

Although primarily a bird of woodland edge and glades, this species also frequents parks, large gardens and the neighbourhood of farm buildings. It is most often found in deciduous or mixed woods but also occurs in open pine woods. It nests in holes, on ledges or among ivy, and will use open-fronted nest boxes or half coconut shells. Spotted flycatchers are widely distributed on the mainland and inner islands. Because they are so dependent on flying insects they are relatively late arrivals in spring, often not appearing until the second half of May.

MALE

Pied Flycatcher

Muscicapa hypoleuca
Gael: Breacan Sgiobalt

Widespread   **May–July**

The male pied flycatcher is the *only small woodland bird with a black back and white underparts*; it also has a *broad white wing patch* and a smallish white area on the forehead. Females are olive-brown above and less pure white below, and they have a smaller wing patch. Both sexes have *white outside edges to the tail*; these and the wing patches show well in flight. After alighting the birds often flick their wings and move their tails up and down. This species differs from the spotted flycatcher in seldom returning to the same perch after a flycatching flight. The pied flycatcher's song is a repeated high two-note 'zee-it', usually concluding with a trilling warble.

This species frequents deciduous woodland, usually near water and where oaks are present, and is most abundant in the Trossachs–Loch Lomond area, Argyll and Dumfries–Kirkcudbright, but occasionally nests further east and north. The numbers breeding in Scotland vary widely between years, as does breeding success; in wet years the caterpillars essential for rearing young are often washed off the trees and so become unavailable to the birds.

FEMALE

Long-tailed Tit

Aegithalos caudatus
Gael: Cailleach Bheag an Earbai

ADULT

Widespread   **All year**

A *tiny body* and *disproportionately long, narrow tail* are striking features of this delightful little bird. The adults' *blackish, pinkish and whitish plumage* is unlike that of any other species; young birds lack the pink and have more extensive black on the head. Long-tailed tits are active and acrobatic, sociable and noisy. They move around in family parties or flocks, keeping up a continual, and easily recognisable, conversational chorus of trilling 'sirrup' and low 'tup' calls. Their long tails are very obvious, both in flight and when moving about in trees, allowing them to be readily identified by silhouette alone.

Long-tailed tits, which build domed nests, are resident in a wide variety of woodland types, but seem to prefer fairly open birch or mixed deciduous woods. They are widely distributed throughout the mainland but absent from many of the islands. In winter they sometimes join up with other tits in roaming flocks, but are often seen in single species groups. They are not regular bird table visitors though they occasionally come to peanut feeders, where they are much less aggressive than other tits.

JUVENILE

See 'Tits' p.40

Crested Tit

Parus cristatus
Gael: Cailleach Bheag a' Chìrein

ADULT

Local   **All year**

This is the *only small bird with a distinct crest*; its *speckled crown feathers* are elongated to a point and can be raised or lowered, making the crest more or less obvious. It also has a distinctive *call, a low purring trill*, rather like that of a long-tailed tit but lower; this serves to keep members of a family party or small group in contact as they move around, busily searching for insects. Crested tits are birds of the pinewoods, in summer feeding mainly in the foliage, but in winter often coming down into the shrub layer of heather and juniper. They are less gregarious than other tits but occasionally join up with coal tits (p.43) and treecreepers (p.111).

Crested tits breed only in the highlands, most regularly in the native pinewoods of the Spey Valley, but also now in the Scots pine plantations of the Culbin Forest. They normally excavate their nesting holes in dead stumps, but have learned to use nest boxes in plantations.

JUVENILE

See 'Tits' p.40

109

Marsh Tit

Parus palustris
Gael: Ceann-Dubhag

Local   **All year**

This tit occurs locally in southeast Scotland, most often in estate policy woodlands. It can be distinguished from a coal tit (p.43) by the lack of a white patch on the nape; from a willow tit by its voice, its tiny and *very neat chin patch*, and the fact that its *black crown is glossy* not sooty; and from a male blackcap (p.101) by its bib, much shorter bill and plumper body. The marsh tit's usual calls are a loud and forceful 'pitchew' and a nasal 'chika-dee-dee-dee', and its song is a repeated 'shippi-shippi'.

Willow Tit

Parus montanus
Gael: Ceann-Dubhag
an t-Seilich

Local   **All year**

The willow tit is locally common only in the west, from the Clyde south to the Solway, where it occurs in scrubby woodland with birch and alder, often along a river or lochshore or on water–logged ground. It has a *larger and less cleanly edged bib than the marsh tit*, is much plumper than a male blackcap (p.101), and lacks the coal tit's (p.43) white nape patch. Its calls are a wheezy 'eez-eez-eez' or a very high 'zee-zee'. Willow tits excavate their own nesting holes, and require dead stumps for successful breeding.

See 'Tits' p.40

Treecreeper

Certhia familiaris
Gael: Snàigear

Widespread   **All year**

This is the *only small bird which feeds by climbing up tree trunks* and branches; it usually spirals up a tree, searching cracks in the bark for insects, then flies down to the base of the next one. Its *rather long downcurved bill* and the stiff pointed feathers in the centre of its graduated tail also distinguish it from all other small brown birds. It is unobtrusive and often first draws attention to itself by its high thin 'tsee' and 'tsit' calls. Its song, a series of 'tsee' notes ending in a faster flourish, is so high-pitched that many people have difficulty in hearing it.

Treecreepers frequent all types of woodland, though they are more abundant in deciduous and mixed woods than in conifer plantations, and are widely distributed virtually everywhere there are trees, except for the less well-wooded islands. In parks and estate woodlands with soft-barked redwood treees, it is easy to spot a treecreeper's roosting place: an oval hollow in the bark, with a white splash of droppings at its lower end. The birds tuck themselves into such sheltered crannies and fluff up their feathers to help them keep warm. They often nest behind a loose slab of bark, or in the crack of a broken branch. They are usually solitary but in winter sometimes join up with parties of tits or goldcrests (p.105).

Redpoll

Carduelis flammea
Gael: Deargan Seilich

MALE

Widespread **All year**

It is often their voice that draws attention to redpolls: an almost trilling or buzzing twitter variously represented as 'tyu-tyu-tyu' or 'zz-chee-chee-chee' and easily recognisable with practice. At close range adults are readily identified by their *small black chin patch and crimson forehead*. Females lack the male's pinkish flush on the breast, and young birds have neither black chin nor red forehead. In flight redpolls show *two rather faint buff wing bars*.

In summer redpolls frequent open birchwoods, willow scrub, or young conifer plantations, and quite often breed in suburban situations. Outside the breeding season they are gregarious, and move around in flocks from one patch of woodland to another, sometimes with siskins. They tend to fly quite high and can be seen only as dots overhead – but their calls enable them to be identified. Many redpolls move south in winter, but some remain, feeding first on birch seeds and later mainly on waterside alders. They are agile and acrobatic and readily hang upside down. This species is the most widespread of the small brown finches, being absent as a breeder only from the Northern Isles and Outer Hebrides.
See 'Finches' p.29

JUVENILE

FEMALE

Siskin

Carduelis spinus
Gael: Gealag Bhuidhe

MALE

Widespread   **All year**

Smaller, neater and *much streakier than the greenfinch* (p.32), this species too has *yellow wing bars and sides to the tail*. Male siskins have a distinctive black cap and chin, and all ages are much more yellow/white below than greenfinches. They are lively and restless, flying in bounding loops and often hanging upside down when feeding. Their usual call is a wheezy 'tsooeet' and their varied twittering song, often ending in a long wheezy note, is given both in display flight over the treetops and when perched.

In summer siskins frequent coniferous woodlands, where they feed on cone seeds, often high up. In winter they move to birchwoods and riverside alders, where they may flock with redpolls, and also come into gardens for peanuts. This is one of the species which has spread as a result of afforestation. It is now widespread over most of the mainland and several of the inner islands, but is scarce as a breeder in parts of the central lowlands and near the east coast. Winter flocks, some of which are of immigrants, visit areas in which siskins do not breed.

See 'Finches' p.29

FEMALE

113

Crossbill

Loxia scotica/curvirostra
Gael: Cam-Ghob

MALE

Widespread   **All year**

Crossbills often first attract attention with their rapidly repeated loud and emphatic 'chip' calls. They feed on cones, usually high in a tree, and sometimes hang parrot-fashion as they wrench off a cone, anchor it with a foot and use their *specially adapted bills* to prise out the seeds. Discarded cones are dropped to the ground; their split and frayed scales distinguish them from those stripped by squirrels. *In flight crossbills look heavy-headed and short-tailed;* they usually fly above the treetops. They move around in family parties or flocks, which usually include birds in various plumages: very red adult males, part-red immature males, olive-green and yellowish females, and heavily streaked brownish-green juveniles.

Native Scottish crossbills breed in pine and spruce woods from Tayside northwards. In some years continental crossbills arrive in large numbers from Scandinavia. With many planted forests now mature and producing good crops of cones, it has been possible for some of these immigrant birds to colonise conifer woodlands in many parts of the country, including areas where Scottish crossbills do not breed.

See 'Finches' p.29

FEMALE

FRESHWATER

Many of the large highland lochs are deep and cold; their waters provide little food for birds other than the saw-billed fish-eating ducks

The birds which are regularly seen on, or close to, freshwater may be using it in a variety of ways. Some spend virtually the whole of their lives on the water, coming on land only to nest: for these the water has to provide all the food they need. Divers, grebes and many of the ducks are in this group. Some are basically land birds which feed in the water: these include such varied species as heron, osprey, kingfisher and dipper. Some, such as pink-footed (p.56), greylag (p.57) and canada geese, and gulls, use lochs and reservoirs only as safe roosting and bathing places. And some, for example swallows (p.44) and wagtails, simply take advantage of the many insects which hatch from the water's surface.

The quantities and kinds of food available to birds vary greatly in different types of freshwater. Shallow lowland lochs usually support a good growth of aquatic plants, and the invertebrates that

FRESHWATER

Fast-flowing streams and rivers, typical of the highlands, are the haunt of dipper and grey wagtail

feed on them, as well as eels and fish such as perch and pike. Deep, cold lochs, in contrast, often have little vegetation but good populations of fish, especially trout. There are contrasts too between different types of running water, but in this case physical characteristics are especially important. Slow-flowing lowland rivers and streams have little turbulence, so fish moving under the water can be readily spotted from above by a fishing bird. Along fast-flowing upland streams and rivers the rapids and falls ensure that the water is well aerated, which encourages the growth of insect larvae. Although the ubiquitous mallard and a few other species can be found on almost any kind of freshwater, many water birds are associated with a particular type.

Water birds are most active early and late in the day, and often have a 'siesta' in the afternoon, when they float with heads tucked round between their wings, apparently sleeping. In midsummer there is a period when many of the ducks 'disappear' while they are moulting; some do move away from their usual lochs but most dabbling ducks simply stay hidden in whatever cover is available near the water's edge. At any time of year it is worthwhile looking carefully along the shoreline for birds that are resting on land. Identifying water birds is easiest when the light is good and the water surface calm, but even in less favourable conditions, when colours are difficult to make out, many species can be recognised by their shape and behaviour.

LOWLAND LOCHS

Shallow lochs surrounded by farmland receive a regular supply of nutrients as fertiliser drains off the land; this encourages a rich growth of vegetation in and around such lochs, while in warm weather an algal growth resembling pea soup often develops at the water surface. The very local shoveler and black-necked grebe are most often seen on these enriched lochs, as are the few breeding pairs of gadwall and pochard. Mute swans, teal, tufted ducks, great crested and little grebes, moorhens and coots are found on a wider variety of lowland lochs, and canada geese at a few localities, including urban parks. All these species may be present at almost any season, as may reed buntings among the surrounding vegetation and herons

A rich growth of vegetation fringes many lowland lochs, both above and below the water, providing shelter, nesting cover and food for a variety of species

stalking their prey at the water's edge. Where there is a lot of floating vegetation there is often a breeding colony of black-headed gulls, while still summer days usually bring swallows, sand martins and swifts (p.46) to hawk for flies over the water, and occasionally common terns (p.196) to fish in the shallows.

Early autumn sees the arrival of wintering wildfowl, such as wigeon (p.199), pochard and goldeneye, and in some areas roosting geese, whooper swans and perhaps a few cormorants (p.186). Flocks of roosting gulls start to gather on many of the larger lochs and reservoirs from late summer, often continuing to use the site until well through the winter, and a few lochs near the east coast are fairly regularly visited by migrating little gulls. In severe winters, when many shallow lochs freeze over, all the birds which feed in the water are forced to move away in search of food, leaving only those which can feed on land, such as geese, mallard and moorhens, to roost on the ice.

HIGHLAND LOCHS

Very large deep lochs such as Loch Tay and Loch Ness, and most hydro-electric dams, support relatively few water birds, often only a few pairs of mallard and saw-billed ducks (p.132-3). The smaller shallower lochs with some reedy areas or clumps of bogbean are of more importance to birds; some of these have breeding wigeon (p.199) or slavonian grebes, while fishing ospreys are frequent visitors to an increasing number. Highland lochs

Lochs set in open moorland are the breeding place of divers; red-throats prefer small lochans and black-throats larger waters

with stony shores often have common sandpipers
and common gulls in summer, and sometimes
redshanks (p.208), oystercatchers (p.204) and
ringed plovers (p.205), nesting on or near the
shoreline while black-throated divers breed on
islands in a few of the larger and remoter lochs.
Peaty moorland pools are the breeding place for
red-throated divers, and provide feeding grounds for
upland breeding birds such as greenshank (p.171)
and dunlin (p.207). In the Outer Hebrides and far

Lowland rivers often have sandy banks suitable for hole-nesting sand
martins and kingfishers, and sometimes shingle islands with breeding
terns, gulls and waders

northwest greylag geese escort their broods onto moorland lochs, and in Shetland a small marshy loch is summer home to most of the very few pairs of red-necked phalaropes nesting in Britain.

RIVERS AND STREAMS

Slow-flowing rivers with sandy banks provide suitable feeding and breeding habitat for kingfishers, and also for sand martins, although the latter occupy other types of site as well. Where there are quiet backwaters mute swans, mallard and moorhens often nest, while in some areas wintering wigeon (p.199) graze on the banks and greylag geese roost there. The larger fish-rich rivers are the main breeding habitat of goosanders, and sometimes also have red-breasted mergansers; in the highlands they are quite often visited by fishing ospreys; and some have small numbers of wintering cormorants. Where there are shingle banks or islands oystercatchers, ringed plovers and common gulls often nest, and occasionally common terns, while faster-flowing rivers, with falls and rapids, and hill streams attract dippers and grey wagtails.

MARSHES AND REEDBEDS

Many low-lying lochs and rivers have adjoining reedbeds or marshy ground, sometimes covering large areas, as at the Insh Marshes, but more often a comparatively small patch. These water-logged areas provide safe havens for water rails and nesting teal, both of which are usually difficult to watch as they prefer to stay hidden in cover. Snipe too frequent marshes, which are also the principal breeding habitat of both sedge warbler and reed bunting. In autumn the larger reedbeds attract roosting sedge warblers and flocks of sand martins and swallows, as they start to migrate southwards; pied wagtails (p.36) and starlings (p.26) also regularly roost among reeds, and in a few areas there are winter roosts of hen harriers.

Shy birds such as teal, water rail and snipe take advantage of the dense cover often available in water-logged marshland

Mute Swan

Cygnus olor
Gael: Eala

ADULT MALE

Widespread   **All year**

The *black knob at the base of the* mute swan's *orange bill*, its *curved neck* when swimming, and its noisy wing beats in flight help to distinguish it from the whooper swan. Mute swans feed on aquatic plants, often using their feet to help them remain up-ended in the water so that they can reach deeper. They breed solitarily on lowland freshwater, beside slow rivers as well as ponds and lochs, and also by sheltered tidal waters. In late summer moulting flocks gather on large lochs and estuaries, mainly in the east but also in South Uist and Orkney.

Whooper Swan

Cygnus cygnus
Gael: Eala Bhàn

ADULT

Widespread   **October–April**

Whooper swans are gregarious and noisy, giving their bugling 'whoop-whoop' calls frequently, both on the ground and in flight. They *hold their necks much straighter* than a mute swan and have *no bill knob*; the *black and yellow bill* pattern varies in different individuals, but the tip is always black. They winter mainly in lowland farming areas, where they roost on lochs and feed over stubble and potato fields as well as in lochs and marshes, though some family parties occupy hill lochs.

Canada Goose

ADULT

Branta canadensis
Gael: Gèadh Chanada

Local   **All year**

This *large, dark* goose was introduced to Scotland many years ago, initially on ornamental and park ponds. It has now become established in widely scattered areas and the population is steadily increasing. Canada geese in this country are not migratory, but from June to August large numbers, many of them from Yorkshire, gather to moult on the Beauly Firth.

See 'Geese' p.55

DUCKS

Ducks
- often have markedly different breeding plumages in male and female
- during summer moult males (in eclipse) lose distinctive colouring and look very like females
- often have distinctive white or coloured bars or patches on their wings
- have differently shaped bills according to the foods they take
- are often separated into groups according to how &/or where they feed:

Shelduck (p.198)
- is almost goose-sized
- feeds over mudflats
- occurs only locally inland

Shelduck

Surface feeding ducks: mallard (p.126), teal (p.127),
  wigeon (p.199), pintail (p.198), gadwall (p.127) &
  shoveler (p.128)
- dabble in shallow water or up-end
- feed mainly on vegetable matter
- spring into flight from the water
- often have a blue or green bar
  (speculum) at the back of the wing
- often rest, and sometimes feed, on land

Mallard

Diving ducks: tufted (p.130),
  goldeneye (p.131), pochard (p.129)
  & scaup (p.202)

Tufted duck

- dive for food, often with a jump before going under
- feed mainly on small water animals
- usually patter along the surface when taking off
- often have conspicuous white wing bars or patches
- come ashore only to nest
- can be confused with grebes (p.135),
  coot (p.139) or moorhen (p.138)

Saw-billed ducks: goosander (p.132) &
  red-breasted merganser (p.133)
- dive for food, often sliding under head first
- feed exclusively on fish
- have long bodies and very long slim bills
- sometimes rest on rocks or
  logs but stay close to water
- can be confused with divers (p.134)

Goosander

Sea ducks: eider (p.200), common & velvet
  scoter (p.201) & long-tailed duck (p.202)
- dive for food
- feed on shellfish and crustaceans
- apart from eider, are unlikely
  to be seen ashore

Eider

Mallard

Anas platyrhynchos
Gael: Tunnag Fhiadhaich

ADULT
MALE

Widespread   **All year**

Apart from the drake's glossy green head, the best distinguishing feature of a mallard is its *broad navy/purple speculum narrowly edged with white*, which is conspicuous in flight and usually visible in the closed wing. When the birds up-end to feed, or stand around on land, their *bright orange legs and feet* are very obvious. It is the females which give the familiar 'quack'.

Mallard breed almost anywhere other than the higher hills, and feed on lochs, mudflats and farmland. In winter they desert upland areas and join large flocks at estuaries and lowland lochs. They are adaptable and often tame, taking a wide variety of foods and living happily close to man. Their breeding season starts early and from February onwards they can be seen chasing around in their fast display flights. When moulting their flight feathers, males in June–August and females a few weeks later, they are often difficult to spot as they spend much of their time amongst reeds, where they are concealed but still benefit from the safety of the water.

See 'Ducks' p.124-5

FEMALE

FEMALE

Teal

Anas crecca
Gael: *Crann-Lach*

MALE

Widespread   **All year**

This *small*, neat duck can be difficult to see, as it spends much of its time in cover. At a distance the drake looks grey with a dark head; in flight the *metallic green speculum, bordered with white*, shows up well in both sexes. The male has a musical and carrying 'krik' call, and the female a high 'quack'. Teal rise almost vertically off the water and fly very fast and erratically.

For breeding, teal need a combination of open water and marsh or damp scrub vegetation; they nest in wetlands from low ground well up into the hills. In winter flocks of several hundred gather on the Solway and some of the east coast estuaries, but most remain, thinly scattered, on lochs and marshes. They do not feed over fields as mallard do.

FEMALE

Gadwall

Anas strepera
Gael: *Lach Ghlas*

MALE

Local   **April–October**

Gadwall are most easily recognised in flight, when the *white bar at the back of the wing* is very obvious. They are scarce summer visitors to Scotland, breeding at Loch Leven and a few places in Fife and Tayside, and also in North Uist. Odd birds are sometimes present in winter.

See 'Ducks' p.124-5   FEMALE

Shoveler

Anas clypeata
Gael: Gob-Leathann

ADULT MALE

Local   **April–November**

Shovelers *feed in a characteristic manner*, swimming with head low and neck stretched forward and sweeping their bills from side to side on the water surface; the bill is specially adapted to filter out tiny particles of food as the water passes through it. They seldom leave the water and when not feeding usually hold their heads 'hunched between their shoulders'. Its *long head and distinctive shovel-shaped bill* give this species a rather unbalanced appearance, and make it look in flight as though its wings are set far back. The male's breeding season plumage is unmistakable, and both sexes show pale *blue shoulder patches* when flying.

Small numbers of shoveler breed regularly on rich lowland lochs in the southwest, the central lowlands, Orkney and the Outer Hebrides, and occasionally elsewhere. This species is most widely scattered and abundant in September–October, when Scottish breeding birds are starting to move south and immigrants arrive from the continent; few winter here, however, most leaving before the first hard frosts occur.

FEMALE

See 'Ducks' p.124-5

Pochard

Aythya ferina
Gael: Lach Mhàsach

ADULT MALE

Widespread   **August–March**

The drake pochard's *chestnut head, grey back, and black breast and tail* are distinctive; female goosanders and red-breasted mergansers (p.132-3) also have chestnut heads but lack the black patches and are a different shape. Pochard ducks can be confused with female tufted ducks (p.130), with which they often associate, but are paler brown and have less rounded heads, with a longer more *sloping forehead*; the sides of the head are usually noticeably lighter than the crown. In flight both sexes show a rather *indistinct greyish wing bar*, whereas the tufted duck's is vivid white. Pochard spend much of the daytime floating around, apparently asleep, and dive most actively at dawn and dusk.

This species is present from late summer to spring on many low-lying lochs and reservoirs, most often in small parties but with flocks of hundreds sometimes present, and even thousands on Loch Harray in Orkney. The first to arrive in this country are usually all drakes, the females appearing a week or two later. Very small numbers breed on a few shallow lochs which have abundant aquatic plantlife and plenty of cover round about.

FEMALE

See 'Ducks' p.124-5

Tufted Duck

Aythya fuligula
Gael: Lach an Sgùmain

ADULT MALE

Widespread   **All year**

This is the commonest and most widely distributed of the resident diving ducks. Drakes in breeding plumage have smart *black and white plumage and a drooping crest*, but *females* and moulting males *are a nondescript dark brown*, with only a faint crest showing at the back of the round head. Some females have a narrow white band round the base of the bill, similar to but smaller than that of a female scaup (p.202). At close range the *yellow eye* can be seen, and the *long white wing bar is very obvious in flight*. This species is usually silent, but its wings make a whistling sound as it flies.

Tufted ducks are most often seen on still freshwaters, sometimes occupying quite small lochs during the breeding season but moving to larger ones in winter, and occasionally resorting to large rivers or estuaries in hard weather. They are most abundant in the lowlands, and regularly frequent park ponds. The numbers on a loch are often quite small but even when large numbers gather in winter they are usually scattered over the water and do not form a close flock.

FEMALE

See 'Ducks' p.124-5

Goldeneye

Bucephala clangula
Gael: Lach a' Chinn Uaine

ADULT MALE

Widespread October–April   **Local in summer**

Goldeneye have *small bills and almost triangular heads*, high at the crown. Males can be distinguished, in all plumages, from drake tufted duck by their white chests, and females by the combination of chocolate brown head, white collar and greyish body, with a white wing patch visible towards the tail. *In flight both sexes show very conspicuous square white patches reaching almost to the front of the wing.* Goldeneye are restless, active and agile, diving frequently and able to rise more rapidly from the water than other diving ducks. They often dive all at the same time, so that a whole group suddenly vanishes from sight.

There is a small but slowly expanding breeding colony on Speyside, and occasional birds summer in other areas, but most goldeneye are winter visitors. Wintering birds are widely though rather thinly scattered over lochs and rivers, but gather in bigger concentrations on the large estuaries, where they feed near sewage outfalls. Breeding birds are usually on small lochs surrounded by woodland; they nest in holes and on Speyside many occupy nest boxes.

FEMALE

See 'Ducks' p.124-5

131

Goosander

Mergus merganser
Gael: Lach Fhiacailleach

ADULT MALE

Widespread   **All year**

The creamy, almost *pinky, white of the breast and sides* of a drake goosander in breeding plumage stands out, even when the light is so poor that the glossy dark-green head is difficult to see. Females can be distinguished from red-breasted mergansers by the *clear-cut junction between chestnut and grey-white on the neck.* In this species the *crest droops down the back of the head* and is often not obvious. Goosanders frequently swim with head submerged before diving for fish; family parties often all dive at the same time. Both they and mergansers are usually silent.

This is much more a bird of freshwater than is the merganser. It breeds on large lochs and fast rivers, mainly in the central highlands and southwest, and is absent from almost all the islands. In winter it is more widely distributed, occurring on many lochs not occupied in the breeding season, especially in the south and east. From May to September moulting goosanders gather on both fresh and tidal waters in a few areas, but the only large wintering flocks are on the Tay and Beauly Firths.

FEMALE

See 'Ducks' p.124-5

Red-breasted Merganser

Mergus serrator
Gael: Sìolta Dhearg

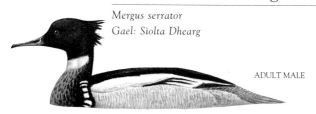

ADULT MALE

Widespread   **All year**

Markedly long and slim bodied and billed, this fish-eating duck can be confused with the similar but larger goosander. Males in breeding plumage are the easiest to distinguish; they are *dark chestnut on the 'chest'* where a goosander is white, and *grey, not white, along the sides of the body*. The most reliable distinguishing features in females are the red-breasted merganser's *ragged wispy crest, which sticks up and out at the back of the head*, and the *gradual merging of chestnut with grey on its neck*. When swimming, mergansers often jerk their heads forward in time with each push of their feet. Females of the two species are difficult to distinguish in flight.

Mergansers breed on freshwater and sea lochs, and on rivers, most abundantly in the northern and western highlands and the islands. They are absent from most lowland areas and scarce in parts of the central highlands. After breeding they move to the coast, where flocks gather to moult before dispersing to winter widely scattered on tidal water. Large numbers of immigrants are also present in the Firths of Forth, Tay and Moray in winter.

See 'Ducks' p.124-5   FEMALE

DIVERS

Divers
- dive by sliding under head first, submerge when alarmed
- fly fast, with head and neck lower than body
- nest on the ground beside open-shored lochs
- winter on the sea
- can be distinguished by bill shape and head posture

Red-throated Diver

Gavia stellata
Gael: Learga Mhòr

Local   **All year**

This diver's *slender, up-tilted bill* is held slightly raised. It nests beside small moorland lochs and flies to the sea to feed, giving a goose-like cackle as it goes; its display involves high wailing calls, often in duet. Red-throated divers breed mainly in the Northern and Western Isles and northwest highlands, with only a few pairs outside this area. From September to March large numbers are off the east coast.

Black-throated Diver

Gavia arctica
Gael: Learga Dhubh

Local   **All year**

The black-throated diver's *bill is straighter, heavier and held more level* than a red-throat's. Large lochs, often with islands, provide both nesting sites and fishing grounds for this species, which breeds in the northwest highlands and western islands but not the Northern Isles. Small groups winter around the Hebrides and Northern Isles and off the east coast.

GREBES

Great crested (p.136), black-necked (p.137), slavonian (p.137) & little (p.135) grebes
- are water birds with sharply pointed bills and a 'tail-less' appearance
- sexes look alike
- surface dive, sliding under or jump-diving
- fly less often than ducks
- breed on freshwater lochs; some winter on coastal inshore waters

nestling great crested grebe

- build floating nests attached to emergent vegetation
- chicks have striped head and neck and are often carried on parent's back
- size (great crested), pattern & colour of head & neck, & shape of body (little), are main distinguishing features

ADULT SUMMER

Little Grebe (or Dabchick)

Tachybaptus ruficollis
Gael: Spàg-ri-Tòn

Widespread   **All year**

This is the *smallest grebe*, with a distinctive *stocky* appearance due to its short neck and *blunt, rounded tail-end*. In summer its brownish colour is relieved by *chestnut cheeks and throat* and a *pale yellowish spot at the base of the bill*. Commoner and more widespread than the other small grebes, it breeds on lochs and ponds with plenty of cover, mainly in the lowlands. Little grebes give a loud, whinnying trill in the breeding season, often calling from cover. They fly more readily than the other small grebes, and often dive with a noticeable splash. Some winter in sheltered coastal waters, when the typical rounded stern and short straight bill help identification.

ADULT WINTER

135

Great Crested Grebe

Podiceps cristatus
Gael: Gobhlachan

ADULT SUMMER

Widespread   **All year**

A short *black crest and black-tipped chestnut ruff* around the top of its *long neck* distinguish this, the *largest, grebe* in summer. In winter its crest is barely detectable and its ruff disappears, leaving it very white-faced. At all seasons its *silky white breast* looks brilliant when it tips sideways to preen. Great crested grebes often float with necks drawn back and bills tucked in. They perform elaborate displays, including head-shaking with raised crests and a 'penguin dance' when the pair, each carrying weed, rise upright on the water, face to face. They give various harsh and guttural growls, croaks and barks.

Great crested grebes breed mainly in the central lowlands, on low lying lochs with enough emergent vegetation to provide cover for their nests and a good supply of eels and small fish. Sudden heavy rainfall often results in nests being washed out. In late summer many move to the coast, where they winter in relatively sheltered waters; there are especially large numbers in the Firth of Forth and Loch Ryan. Return to the breeding lochs may take place as early as February in open winters.

See 'Grebes' p.135

ADULT WINTER

Slavonian Grebe

Podiceps auritus
Gael: Gobhlachan Or-Chluasach

ADULT SUMMER

Local   **All year**

The conspicuous *golden 'shaving brush', extending from in front of the eye to the back of the head*, where it forms short 'horns', and its *chestnut neck, upper breast and flanks* distinguish this species from the other *small* grebes in summer. During the breeding season, April to October, it occurs locally on highland lochs with some emergent vegetation, mainly in Inverness-shire. In winter, when small parties may be seen on inshore coastal waters, it looks more cleanly patterned than little (p.135) or black-necked, with its black crown sharply separated at eye level from white cheeks and its neck white except for a dark line down the back.

Black-necked Grebe

Podiceps nigricollis
Gael: Gobhlachan na h-Amhaiche Duibhe

ADULT SUMMER

Local   **Mainly April–September**

The *black neck and upper breast*, steep black forehead and high crown, and *drooping fan of golden feathers behind the eye* are distinctive in summer, when this grebe is present on a few rich lowground lochs in the central lowlands. Small numbers winter around the coast, especially in Loch Ryan. Outside the breeding season the black-necked looks dingier than a slavonian grebe, with cheeks and neck smudged greyish, while its longer body and fine up-tilted bill distinguish it from the little grebe.

See 'Grebes' p.135

Moorhen

Gallinula chloropus
Gael: Cearc-Uisge

ADULT

Widespread   **All year**

Its *red bill and forehead*, white line along the side, and *white patch under the cocked tail*, distinguish the moorhen from other plump, dark water birds. As it swims it nods its head and jerks its tail; it patters along the surface when taking flight, and flies low and only for short distances. Young birds are brown and do not have red bills. A liquid 'prruk' is the most distinctive note, but calls include a variety of chattering and squeaking sounds.

Moorhens feed mainly at the water surface, only occasionally up-ending or diving, but also visit fields, especially in winter. They nest beside ditches and slow rivers as well as lochs and ponds, and are widespread in the lowlands but scarce over much of the highlands and on most of the islands. Although they look ungainly on land, lifting their feet high as they walk, they are able to climb into bushes, where they sometimes place their nests, and perch when the water is frozen. Moorhens operate an 'extended family' system, with young from the first brood helping to feed chicks from a later hatching.

ADULT

Coot

Fulica atra
Gael: Lach a' Bhlàir

ADULT

Widespread   **All year**

The *white bill and forehead* identify the *bulky* black coot, which has a tail-less look as its *rounded back slopes down at the rear*. Young birds are grey with whitish bellies. The commonest call is an explosive 'kut', from which the bird gets its name. Coots feed on underwater plants, diving for them with an audible 'plop' or scavenging vegetation pulled up by swans or ducks. Their flight is laboured, with a long surface-pattering take off and a breast-first splash landing. On land their gait is a waddling run, often with flapping wings.

Coots are gregarious, noisy and aggressive, constantly indulging in territorial squabbles, with birds charging at each other with raised wings and lowered heads. They are seldom seen alone and often form large flocks, especially in winter when many immigrants are present. Although they too sometimes forage over farmland, they are more dependent than moorhens upon still water and occur mainly on large lochs and reservoirs on low ground. Coots are scarce in the highlands and islands.

JUVENILE

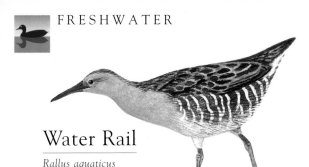

Water Rail

Rallus aquaticus
Gael: Snagan Allt

Widespread **All year**

Water rails are *seldom visible*, spending most of their time skulking in reedbeds and similar dense marshy vegetation. They are *very audible*, however, emitting, often at night, an *astonishing range of squeals, grunts and screams* reminiscent of pigs squabbling for food. In summer they are widely but thinly scattered wherever there are extensive areas of suitable wetland. In winter, when immigrants arrive, they occur also in ditches and are sometimes seen along river banks during hard frost.

Red-necked Phalarope

Phalaropus lobatus
Gael: Deargan Allt

FEMALE

Local **Late May–early September**

This *very small* wader is one of Britain's scarcest breeding birds. There is a small colony in Shetland, on the RSPB's Fetlar Reserve, but otherwise only a very few nesting pairs. Phalaropes summer on lochs and marshes with small pools, and winter at sea. They *swim high in the water, twirling round* as they grab insects on the surface, and come onto land only at the nest. They are unusual in that the female is more brightly coloured than the male, which is responsible for incubating the eggs and rearing the chicks.

140

Osprey

Pandion haliaetus
Gael: Iolair-Iasgaich

ADULT

Widespread   **Late March–late August**

In flight the *dark-above, white-below* osprey looks a bit like a black-backed gull (p.190), but its *wings* are longer and less pointed and *have a pronounced angle at the 'wrist'*. Seen from below it has a similar pattern to a buzzard (p.163), but longer and narrower wings. Much of the male's time is spent sitting upright, high on a dead tree, keeping watch. Ospreys fish most actively early in the day, splashing feet first into the water and rising with their catch firmly gripped in their talons. Their call, usually given near the nest, is a short whistling 'pew'.

Most Scottish ospreys arrive from their African wintering grounds in early April and occupy a territory which includes large trees, often pines, for nesting, and good fishing in lochs or rivers. The same nests are used year after year, with fresh sticks and lining added annually. More pairs nest each year, and breeding ospreys are now quite widely scattered from the central lowlands northwards. June and July are the best months for watching young birds from the hides at Loch of the Lowes and Loch Garten.

ADULT

See 'Birds of prey' p.161

Common Gull

Larus canus
Gael: Faoileann

ADULT

Widespread   **All year**

This gull is *similar* in plumage pattern *to a herring gull* (p.191) *but* is *much smaller* and has *no red spot on its greenish-yellow bill*. It can be distinguished from the slightly smaller black-headed gull at all seasons by the absence of any coloured feathering on its head. Its call is a squealing 'keeya', rather high and shrill.

In summer common gulls are widely distributed, nesting colonially on rocky islands, on shingle banks along coasts and rivers, on moorland bogs, and beside hill lochs with stony shores. In winter they desert the moorlands and move to farmland and estuaries, roosting on nearby lochs and reservoirs or sandy beaches. In late summer and autumn large numbers of immigrants from the continent arrive and by mid-winter there are flocks of several thousand at various places near the east coast. When feeding, common gulls run behind the plough, make short leap-frog flights as they hunt for worms and insects on grassland, and pick prey items off the surface of water. They are much warier than black-headed gulls and do not gather at picnic places or visit gardens.

See 'Small gulls' p.195

ADULT

Black-headed Gull

Larus ridibundus
Gael: Faoileag Dhubh-Cheannach

ADULT SUMMER

Widespread   **All year**

This is the least sea-frequenting of the gulls and by far
the tamest and the commonest inland. Its *red bill
and legs*, and in summer its *chocolate-brown head*,
are distinctive; in winter its head is white
with smudgy dark spots in front of and
behind the eye. In flight the *white front edge to
the wing* makes it look paler than a common gull.

FIRST
WINTER

Newly fledged young are a scaly red-buff above,
and first-winter birds have darkish bars on their
wings. Black-headed gulls are gregarious and noisy
all year; their usual call is a harsh scolding 'karr'.

This species breeds colonially on lowland bogs, marshy
loch edges, and sand dunes, always near shallow calm water;
it often nests among floating vegetation and the chicks have
to swim almost as soon as they hatch. Black-headed gulls
winter on farmland, around estuaries and near towns, readily
coming to parks and gardens, and scavenging at
picnic areas and other places where food is made
easily available. They can often be seen
hawking over grassland for flying insects
such as craneflies and ants.

See 'Small gulls' p.195

ADULT WINTER

ADULT SUMMER

Little Gull

Larus minutus
Gael: Crann-Fhaoileag

Local **March–April & July–October**

Small parties of little gulls visit the east coast on migration. They are most regularly seen from Angus south, hawking for insects over freshwater lochs, less often on the coast itself or adjoining moorland. Adults look rather like black-headed gulls (p.143) but have *jet black*, not brown, heads, *no black on the wings*, and *dark underwings*. Blackish zigzags on the back make young birds resemble young kittiwakes (p.191), which are not normally found in the same habitat.

ADULT SUMMER

Sand Martin

Riparia riparia
Gael: Gobhlan Gainmhich

Widespread **Late March–September**

This is the *only small brown and white bird which regularly feeds by hawking over water for flying insects*. Sand martins are very gregarious and are usually seen in groups, when they keep up a rather harsh conversational twittering. Their flight is fast and agile, with frequent flutterings and changes of direction. They nest colonially in exposed vertical faces of sand or fine gravel which are either over water or relatively high; when a bank becomes vegetated, or slips from the vertical, the site is abandoned. Before leaving in late summer sand martins gather in large flocks.

Grey Heron

Ardea cinerea
Gael: Corra-Ghritheach

ADULT

Widespread   **All year**

The heron's *long neck and legs* are distinctive when extended, but it looks quite different when standing with its head hunched between its shoulders and its neck virtually invisible. *In flight its head is always drawn back*, so that it appears to have a short thick neck, and its *legs project beyond its tail*. A harsh 'kraaak' call is often given in flight, while nesting birds produce a variety of bill-snapping, croaking and other noises.

Herons are widely distributed and catholic in their habitat tastes. In many areas they are associated with freshwater, feeding in marshes and shallow lochs or on damp farmland, but on the west coast they spend much of their time along the shore, fishing in tidal water. They feed solitarily but nest in small colonies, usually in woodland and quite often in conifer plantations. The tidal-feeding west coast population is unaffected by severe winters, whereas in the colder east numbers decline, due to the herons' difficulty in finding food when both freshwater and ground are frozen.

ADULT

Snipe

Gallinago gallinago
Gael: *Naosg*

Widespread   **All year**

Snipe are *more likely to be seen in flight than on the ground*, where their *camouflage colouring* and secretive behaviour make them difficult to spot. In spring and early summer their display flights, often performed at dusk, attract attention: these involve 'chip-per, chip-per' calls and oblique downward dives accompanied by a loud, almost buzzing, drumming noise, produced by the passage of air through the spread tail. *When flushed snipe give a hoarse, rasping 'skaap' and fly off in a series of erratic zigzags*, before dropping into cover again. The *very long slim bill* is always obvious.

Snipe require soft, marshy ground to feed in, as they probe deep into mud. They breed on boggy moorland, in marshes and on damp rough grazing with clumps of coarse vegetation. In winter they become scarce inland, as frozen ground makes feeding impossible. Many Scottish-breeding snipe move to Ireland, where the climate is milder, while immigrant birds arrive from the north and east. Although quite large numbers may be present on a wetland in autumn the birds remain solitary.

See 'Waders' p.203

Common Sandpiper

Actitis hypoleucos
Gael: Luatharan

Widespread   **May–August**

This is the only wader closely associated with upland streams and stony rivers and loch shores. Apart from a *conspicuous white wing bar*, it has no distinctive markings but its *behaviour is characteristic*. It flies low, with fluttering wing beats alternating with glides, wings held below body level, and when standing *constantly bobs its head and body and wags its tail*. When flushed it gives a shrill, ringing 'swee-wee-wee' call as it flies away above the water.

Common sandpipers breed wherever there is suitable habitat, well spaced out along a river or loch shore. When they desert the nesting area in late summer they move first to mud or shingle shores on the coast, and later leave Britain altogether to winter in Africa. Unlike many other waders they are not at all gregarious, even in winter; although quite a number may be present on an estuary they remain scattered around, and do not feed or roost together in a flock.

See 'Waders' p.203

Kingfisher

Alcedo atthis
Gael: Biorra-Crùidein

ADULT

Widespread   **All year**

Usually seen as a *vivid flash of blue* with whirring wings as it heads away low and fast over the water, the kingfisher is unlikely to be confused with any other species; when seen side on in flight or perched its *chestnut underparts* are very striking. It dives head first when fishing, plunging in either from a perch, where it may bob nervously as it watches the water for prey, or after hovering briefly in flight. It often returns to the same perch between dives, and usually batters larger fish against the perch before swallowing them.

A bird of slow-flowing lowland rivers, the kingfisher is thinly scattered as far north as the Moray Firth. Its distribution is largely determined by the availability of small fish, and it sometimes takes advantage of fish farms to obtain a guaranteed food supply! Kingfishers suffer badly in severe weather, when rivers freeze, and a few hard winters in succession greatly reduce the population. The fact that two broods are raised in good years helps the species to recover again quite quickly after such a setback.

ADULT

Dipper

Cinclus cinclus
Gael: Gobha-Dubh

Widespread   **All year**

Its *dumpy* shape, brilliant *white shirt front*, and habit of
crouching and bobbing its whole body up and down, make
the dipper easy to identify. It is a bird of fast-flowing
streams and rivers, and also frequents stony loch shores in
winter. It feeds mainly on the stream bed, either walking in
from a stone or plunging into the water, and always facing
into the current. Dippers fly low and fast above the water,
swim buoyantly and sometimes dive under the ice. Their
call is a sharp 'zit' and they sing their high, rather grating,
warble all year.

This species is widespread on the mainland and most of
the inner islands but absent from the Northern Isles and
scarce in the Outer Hebrides. Dippers are highly territorial
and widely spaced out along suitable stretches of river.
They tend to use traditional nest sites, which are often
under bridges or on
a rock; the same
domed nest may be used
in successive years or a new
one may be built nearby.
Although usually solitary, dippers   JUVENILE
sometimes roost in small
groups during the winter.

Grey Wagtail

Motacilla cinerea
Gael: Breacan-Baintighearna

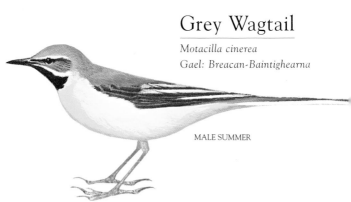

MALE SUMMER

Widespread   **March–August**

The grey wagtail is typical of hill streams and fast rocky lowland rivers, a quite different habitat from that of the yellow wagtail (p.70), which also has a *yellow breast*. The grey is much more *contrasty in colour* and is longer and slimmer-looking due to its *very long tail*, which is constantly whipped up and down. It is restless and active, running over stones to pick up insects and flycatching, from a perch or the ground, with fluttering leaps and hovers. Its direct flight is bounding and it often gives its shrill 'siz-eet' as it flies.

In summer grey wagtails are widely distributed on the mainland and most of the large inner islands but scarce in the Outer Hebrides and Northern Isles. They are usually quite thinly scattered along suitable stretches of water and are most abundant where there are rapids or similar areas of broken water. In winter most move no further than England and a few remain in the central lowlands; sudden periods of severe frost cause heavy mortality, after which the breeding population is much reduced.

JUVENILE

Reed Bunting

Emberiza schoeniclus
Gael: Gealag Dhubh-Cheannach

MALE SUMMER

Widespread   **All year**

The *black head and throat, with white moustache*, immediately identify the male reed bunting in summer; although less dark and distinct, the same pattern is also evident in winter. The *streaky-brown females* and young birds are rather nondescript, but can be distinguished from female yellowhammers (p.74), which also have *white outer tail feathers* and a similar face pattern, by the lack of chestnut on the rump and of any yellowish tinge in the plumage. The reed bunting's song is a monotonous, deliberate 'tsee-tsee-tsee-tissick', dropping towards the end; it often sings perched on a tall reed.

Reed buntings are widespread wherever there is marshy ground or reedbeds, on the mainland and many of the islands. In some areas they also frequent hedges, especially in winter, and young conifer plantations. They are more solitary than most other buntings and finches but in winter sometimes join up with yellowhammers or mixed flocks feeding on farmland, especially stubble fields.

FEMALE

See 'Buntings' p.29

Sedge Warbler

Acrocephalus schoenobaenus
Gael: Glas-Eun

ADULT

Widespread   **Late April–August**

This *skulking, streaky-brown* warbler *sings loudly and vigorously*, with a mixture of musical and harsh, chattering phrases, and *often mimics other species*. It sings day and night, and is consequently sometimes mistaken for a nightingale (which only rarely occurs in Scotland). It shows an unstreaked tawny rump in flight, and its *creamy eye stripe* is conspicuous when it perches in the open, usually while singing. It also sings in a short, vertical display flight. The sedge warbler's calls include various harsh 'churrs' and 'chuks'. It spends much of its time in thick cover, where it climbs agilely as it hunts for insects.

Sedge warblers are widely distributed in the southern half of the country but are rather local further north and absent from Shetland and the Outer Hebrides. They usually breed near water, in reedbeds, beside overgrown ditches or among willow thickets, and occasionally in young conifer plantations at the early thicket stage.

See 'Warblers' p.97

HEATH & HILL

Grassy, sheep-grazed, hills offer little variety of food or nesting sites and consequently support only a few bird species

A very large proportion of Scotland's countryside can be described as heath or hill, terms which include rather varied habitat types all of which share certain characteristics: they are largely open ground, rough and uncultivated, with no more than scattered bushes and few if any trees. Within this general group come 'heaths' such as golf courses and other gorse-dotted rough ground as well as heather moorland, and 'hills' ranging from the rounded grassy hills common in the Southern

Uplands, to the bare granite plateaux of the Cairngorms, and the steep scree-sloped mountains of the north-west. Many of these areas provide only limited resources for feeding and breeding birds: they support a restricted range of plants and consequently relatively few insects, and offer suitable nesting sites for only ground or crag nesting species. However, although their bird populations are sparse and often include only a handful of species, these extensive heaths and hills are the haunt of some of our most exciting birds, among them most of the larger birds of prey.

In Scotland heather-covered ground can be found from near sea level to high up on the hills, and a few bird species, such as dunlin and meadow pipit, are equally at home at either altitude. Many, however, show a distinct preference for a particular type of country, for example the rolling grouse moors, pool-dotted boggy peatlands, or bare high tops. Many of the heath and hill birds can be seen on ground to which the public has unrestricted access, for example on National Trust for Scotland mountain properties or around the chairlifts at Cairngorm and Glenshee. Elsewhere it is advisable to stay on right of way tracks or obtain permission before going onto hills or heather moorland, especially from 12 August until December when shooting of grouse or deer may be in progress. When planning a visit to the hills it is important to remember that the Scottish climate is very unpredictable: winds can rise and temperatures drop within a surprisingly short time, so always take warm and waterproof clothing and some food, wear strong boots or shoes, and carry a compass and the relevant map.

Rough grassland with bushes attracts small upland birds such as linnet, whinchat and stonechat; ground-nesters like curlew and meadow pipit nest in the more open areas

LOW-GROUND HEATHS AND GRASSY HILLS

This type of country, much of which lies south of the highlands, is essentially rough grassland, sometimes with bracken-covered areas or scattered bushes; it is generally grazed by sheep. The most typical birds of the open grassland are curlew (p.62), meadow pipit and skylark, with wheatears where there are stone dykes or rocks and loose stones breaking up the surface of the turf. Where there are clumps of gorse or other bushes linnets, whinchats and stonechats are likely to be present, while cuckoos tend to move freely between heathy ground and nearby open woodland. Buzzards, kestrels (p.67), hen harriers and short-eared owls sometimes hunt over these rough grasslands, though the first two only do so if they are within easy range of a suitable tree or cliff nesting site. Fast-flowing streams in any type of hill country usually attract dippers (p.149) and grey wagtails (p.150).

Grouse moors, with their typical patchwork of colour, are summer home to golden plover and merlin as well as red grouse

HEATHERY MOORLANDS AND
BOGGY PEATLAND

To most people heather moorland implies grouse moor: wide areas of gently undulating country, usually with a distinct mosaic of different colours showing where burning has been carried out in successive years to ensure that there are always plenty young shoots for red grouse – and sheep – to feed on. Such moors are important breeding grounds for hen harriers, merlins, golden plovers, short-eared owls and sometimes dunlins (p.207), but moorland of this type is becoming scarcer as a result of afforestation, which immediately forces out the waders and later the grouse and owls. Where the moors are broken up by rocky gullies, perhaps with a few stunted trees, ring ouzels are often present, as well as wrens and occasionally willow warblers. And where there is woodland nearby there are likely to be black grouse, which in spring can sometimes be seen displaying at the moor's edge. Sheep-grazed moors are often frequented by carrion/hooded crows (p.65) and ravens, which scavenge on the carcases of casualties.

It is in the far north that heather moorlands come closest to the sea, and these rather bare and exposed moors near sea level have their own distinct breeding populations. Colonies of arctic terns (p.196) frequently nest on such ground, as do both great and arctic skuas (p.187), all within easy reach of their marine food resources, and a few Shetland moors have breeding whimbrels. Where boggy areas and peaty lochans break up the surface of highland moors there are likely to be greenshanks and dunlins, as well as curlews and

The wild and remote moorlands and bogs of the northwest highlands are only thinly populated by birds, but among those that breed there are greenshank and dunlin

redshanks (p.208). Teal (p.127) and wigeon (p.199), and in a few places greylag geese (p.57), breed near moorland lochans, while many small lochs, especially in the Northern and Western Isles, hold red-throated divers (p.134).

THE HIGHER HILLS

The formation, as well as the height, of mountainous areas influences the birds which live among them: steep-sided hills with scree slopes usually support little vegetation and hold fewer birds than more gently rounded ranges well covered with grass or heather. High altitude is really significant for only two species: the ptarmigan, which nests mainly above 800m, and the dotterel, which breeds on high stony plateaux, mainly in the central Grampians. But others, such

Cliff-nesting birds of the highland glens include the golden eagle and the peregrine

as golden plover, dunlin, skylark and meadow pipit, quite often nest high up too. Cliffs and glens among the hills provide suitable nesting sites for golden eagles, peregrines and ravens, all of which hunt – or scavenge – over wide areas of hill country. To spot them one must constantly scan the sky, alert for the glimpse of movement along a skyline which draws attention to a hunting eagle or a family of ravens. It is often possible to watch these birds from a glen road or hill track, without the need to make a major expedition into the hills. One of the scarcest Scottish breeding birds is harder to see: the few pairs of snow buntings (p.212) which nest in this country do so near the summits of the highest tops in the Cairngorms – but in winter this species can often be watched at much lower levels, for example near the chairlift car parks.

Dotterel breed on the rounded plateaux of the central highlands, where the highest corries also hold the few pairs of snow buntings which breed in Scotland

The larger birds of prey are often seen overhead, when wing shape is a useful pointer to identification:

Golden eagle

- very large, wings oblong with 'fingers' – golden (p.162) & white-tailed (p.185) eagles

Osprey

- large, wings with obvious angle at 'wrist' – osprey (p.141) & red kite (p.66)

Buzzard

- large, wings broad & rounded – buzzard (p.163)
- large, wings relatively long for width, held straight – hen harrier (p.164)
- large, wings sharply pointed – peregrine (p.165)   *peregrine*

Some birds of prey perch during daytime in typical situations:
- on telegraph & similar poles or fence posts – buzzard & kestrel (p.67)
- on top or branch of live or dead tree – osprey & kestrel   *Sparrowhawk*
- on garden fences, roofs and among branches of trees – sparrowhawk (p.86)

Kestrel (hovering)

With small birds of prey flying style is helpful for identification:
- hovering over open ground – kestrel
- flying low and fast over open moorland – merlin (p.165)
- flying low & fast close to hedges or through woodland – sparrowhawk

Golden Eagle

Aquila chrysaetos
Gael: Iolaire

ADULT

Widespread   **All year**

Golden eagles are usually seen high overhead, indeed one is most likely to spot them by scanning the sky and hilltops. Size is not easy to gauge at a distance, but the eagle's *flight silhouette is different from that of the much smaller buzzard: the wings are longer and less rounded, the head projects further*, and the *tail is longer and less spread*. An eagle hunting along a hillside flies low, with deep and powerful wing beats, and strikes its prey with its feet while still moving fast. Eagles are usually silent.

Golden eagles are widely scattered over the highlands and larger western islands. They hunt over open upland country ranging from heather moorland to mountaintops. They breed most successfully where there are plenty of mountain hares, young deer, grouse and ptarmigan, and raise fewer young where they are largely dependent upon carrion. They are very early nesters and pairs can be seen soaring in display flight on fine days in mid-winter. Eagles are sensitive to disturbance so suspected or known nesting sites should not be approached in February – March, when eggs can quickly become chilled.

JUVENILE

See 'Birds of prey' p.161

Buzzard

Buteo buteo
Gael: Clamhan

ADULT (PALE)

Widespread   **All year**

Buzzards *look 'neckless' when soaring* with wings held in a shallow V and *rounded tail widely spread*; in direct flight rather shallow wing beats alternate with short glides. When perched – often on a roadside pole, a site never used by an eagle – they *sit upright and have a bulky appearance*; their *bright yellow lower legs are usually visible*. Buzzards call frequently, with a plaintive far-carrying 'mee-ooo' which is distinctive. They vary considerably in colour, some birds being much paler or darker than the usual darkish brown.

This is the commonest large bird of prey in areas where there is a mixture of rough open ground and scattered woodland, especially on fairly low hilly ground. A tree or cliff nester, it is scarce in arable areas and the Outer Hebrides and does not breed in the Northern Isles. Buzzards hunt by pouncing, from a low hover or a perch, on rabbits and other small animals; they also take carrion. In the past they were much persecuted, as representing a threat to game birds, and despite legal protection are still shot or poisoned in some areas.

See 'Birds of prey' p.161

Hen Harrier

Circus cyaneus
Gael: Clamhan nan Cearc

ADULT FEMALE

Widespread   **All year**

Harriers are *slim-bodied* birds which *fly low and buoyantly*. At a distance the male looks almost like a herring gull (p.191) in flight, but has rounder wing-tips, a *white rump* and a grey tail. The larger female's white rump contrasts more conspicuously with the streaky-brown of back and tail. Young birds of both sexes resemble females. Adults give a high chattering 'kek-kek-kek' near the nest and often dive at intruders.

The hen harrier is a bird of heather moorland and very young conifer plantations, breeding in widely scattered areas of the mainland from Galloway to Caithness, and in Orkney and some of the larger western islands. In winter small parties roost communally in marshy areas and hunt over lower ground. When quartering the ground in search of small rodents and birds, hen harriers frequently glide with wings held in a shallow V. Their display flight involves switchbacks, rolls and the aerial passing of food from male to female. They are unpopular with grouse moor keepers, and are sometimes illegally persecuted, on the grounds that they not only take young grouse but also upset shoots by scaring adult birds.

ADULT MALE

See 'Birds of prey' p.161

Peregrine

Falco peregrinus
Gael: Seabhag

ADULT

Widespread   **All year**

The crow-sized peregrine hunts in a very different manner from a buzzard (p.163) or hen harrier, flying high to spot prey and *stooping steeply and at great speed* on the chosen victim. In normal flight peregrines alternate fast wing beats and glides, and sometimes hover. This is the most *compactly built* of the three species, and has *sharply-pointed wings* and a tail that tapers towards the tip. Young birds are browner than adults and streaked below, instead of barred.

A cliff-nester, this species is widespread in mountainous areas and also breeds occasionally on the coast. When disturbed near the nest it gives a harsh, screaming chatter. In winter peregrines hunt over coastal marshes, where they often cause panic among flocks of wildfowl and waders.

JUVENILE

Merlin

Falco columbarius
Gael: Mèirneal

Widespread   **All year**

This *small dark* falcon *flies fast and low*, with rapid turns and climbs. Its *pointed wings* distinguish it from the sparrowhawk (p.86) and its colouring and flight from the kestrel (p.67). In summer merlins are thinly scattered on heather moorland in many districts from Shetland to the Borders and Galloway. In winter they move to lower ground and are quite often seen near the coast, chasing small birds.

ADULT MALE

See 'Birds of prey' p.161

FEMALE

GROUSE

Grouse are large ground-nesting game birds, which have different habitat preferences:
- ptarmigan (p.167) are confined to hills, nesting mainly above 800 metres
- red grouse (p.167) favour open heather moorland
- black grouse (p.166) prefer a mixture of woodland and open moorland
- capercaillie (p.85) are birds of mature conifer woodland

Where these habitats meet, and in winter when all except ptarmigan sometimes visit farmland, there is a possibility of confusing females, but
- ptarmigan always have white wings
- red grouse have dark wings & dark, rather square tails
- female black grouse have barred wings & barred, shallowly notched tails
- female capercaillie have darkish wings & barred, broadly fan tails

ADULT MALE

Black Grouse

Tetrao tetrix
Gael: *Coileach-Dubh, Liath-Chearc*

Widespread   **All year**

Black grouse are most obvious in spring, when gathered at their breeding 'leks' which are usually on open ground near woodland. There the *males display with wings lowered and tails raised to show the white underside*, to a chorus of bubbling musical 'roo-koo'ings, which carry a long way. The females stand around nearby. In winter black grouse often sit in trees. They are widespread in the southern uplands and much of the highlands but scarce north of the Great Glen and absent from most islands.

FEMALE

Red Grouse

Lagopus lagopus
Gael: *Coileach-Fraoich, Cearc-Fhraoich*

MALE

Widespread   **All year**

Its *red comb* and angry breeding season '*go-bak, go-bak*' *calls* make the male easy to identify, especially when it stands on a prominent rock. The well-camouflaged females are less obvious, as they tend to skulk in the heather. When flushed, red grouse take off with an explosive burst of whirring wing beats, then glide with down-curved wings. Grouse are found mainly on heather moors but sometimes visit nearby farmland in winter. They are widespread in suitable habitat, but have decreased in some areas due to moorland afforestation.

FEMALE

Ptarmigan

Lagopus mutus
Gael: *Tàrmachan*

MALE SUMMER

Widespread   **All year**

Its *white wings* identify the ptarmigan at all seasons. In winter both sexes are almost completely white; in summer they are mottled brownish, with varying amounts of white below, and can be difficult to spot as they crouch and 'vanish' against a similarly-mottled background. When flushed they give a rattling croak. Ptarmigan breed on mountain heaths, mainly above 800m in the central highlands but at lower levels in the northwest. They can often be seen near the top of the Cairngorm and Glenshee chairlifts.

See 'Grouse' p.166   FEMALE SUMMER

Raven

Corvus corax
Gael: Fitheach

Widespread   **All year**

Ravens call often in flight and their *voice is distinctive*, a throaty 'pruk-pruk', deeper and less harsh than the 'kraa' of the much smaller carrion crow (p.65). Their flight is direct and powerful, and they frequently *perform aerial acrobatics*, rolling, gliding and diving steeply with folded wings; the wedge-shaped tail tip shows well in flight. They breed early and remain in family parties through the summer, sometimes joining up to form larger flocks in winter.

Ravens frequent hill country, moorland or seacliffs, feeding mainly on carrion; where the supply of carrion decreases in a traditional raven breeding area, as sheep are removed to allow afforestation, the birds often cease to breed and may desert the area altogether. Although generally found in thinly inhabited parts of the country they choose to come close to man on some of the islands, where they scavenge on garbage tips. They are commonest in the west and northwest highlands and the islands, more thinly scattered in other upland areas.

Short-eared Owl

Asio flammeus
Gael: Comhachag Chluasach

Widespread   **All year**

Short-eared owls *hunt by day*, quartering the ground with wavering, *moth-like wing beats* and short glides with wings upraised; when circling in display flight they frequently clap their wings together below their bodies and 'sing' with a deep 'boo-booboo'. They *perch in a crouching position*, on the ground, a rock or a fence post, and at close range look fierce, due to their *glaring yellow eyes*; their short eartufts are seldom visible.

Short-eared owls breed on moorland and recently afforested ground, where they feed on small rodents; breeding success is closely linked to the size of the vole population. Like many predatory birds, they start to incubate when the first egg is laid, so the chicks hatch out at intervals. In a good year for short-tailed voles all the young owls are likely to be reared; in a bad year only the oldest one or two chicks will get enough food to survive. This is the only owl to breed regularly in Orkney and the Outer Hebrides. In winter many short-eared owls move south; those that remain hunt over lower ground and coastal areas, and sometimes roost communally.

Golden Plover

Pluvialis apricaria
Gael: Feadag

ADULT SUMMER

Widespread   All year

This species' *liquid, 'tlooee' calls and rippling trills* are among the most typical sounds of heather moorland and bog in summer, and draw attention to golden plovers when their *spangled plumage* makes them difficult to spot. They breed wherever there is suitable habitat on mainland and islands, and are inland from March to July. In late summer they move towards the coast, where they winter on farmland and mudflats, often resting with lapwings (p.63) on grass or potato fields. When a mixed party is disturbed their fast, direct flight and pointed wings make it easy to pick out the golden plovers. The *black chest and belly* are lost in winter, when there might be confusion with knot (p.206) which feed in tighter flocks and have pale rumps and wing bars.

WIN'

Dotterel

Charadrius morinellus
Gael: Amadan-Mòintich

ADULT

Local   May–July

Dotterel breed on bare mountain plateaux, mainly in the Cairngorms and Grampians but very locally in a few other areas. Males, which incubate the eggs and rear the young, are duller in colour than females. Dotterel are *inconspicuous and very tame*, often allowing a close approach; when flushed they give a quiet trill.

ADULT

See 'Waders' p.203

Greenshank

Tringa nebularia
Gael: Deoch-Bhiugh

Widespread   **Mainly March–October**

The greenshank breeds on damp moorland with small pools and lochans, mainly in the northwest, more rarely the central, highlands and some of the islands. It sometimes feeds by the shores of larger lochs, and when moving south in July–October regularly appears in small numbers on estuaries, where a few winter. It can be distinguished from the rather similar redshank (p.208) by the *absence of wing bars*, and by its *call, a deliberate ringing trisyllabic 'tyu-tyu-tyu'*, usually given in flight.

Whimbrel

Numenius phaeopus
Gael: Eun Bealltainn

Local   **April–August**

The whimbrel is much *smaller than a curlew* (p.62), and has a proportionately shorter bill and a *conspicuously striped crown*. It is *most easily recognised by its call*, seven or more notes all at the same pitch and evenly spaced; this is sufficiently distinctive to allow birds flying overhead in the mist to be identified. Whimbrel breed regularly in Shetland, on heather or grassy moorland, and occasionally at a few other places in the north and the Outer Hebrides. They occur in larger numbers as passage migrants, in May and July–August.

See 'Waders' p.203

171

Cuckoo

Cuculus canorus
Gael: Cuthag

ADULT
MALE

**Widespread
April–early August**
Although the cuckoo's
two-syllable call is
familiar, the bird itself is
often unrecognised. It flies low and fast, usually escorted by
several small birds, and looks rather like a sparrowhawk
(p.86) but has more pointed wings and *white tips to
its long tail*. It perches almost horizontally on
posts or branches, more upright on
wires. Cuckoos breed on moorland
and heath and near open woodland. They are
widespread, but most abundant and obvious in
the northwest and the islands. In Scotland they usually
lay their eggs in the nests of meadow pipits (p.173).

JUVENILE

Ring Ouzel

Turdus torquatus
Gael: Dubh-Chreige

**Widespread
April–September**
Widely but thinly
distributed in most
moorland and mountainous
areas of the mainland
and some of the inner

ADULT MALE

islands, ring ouzels breed on open ground with scattered
trees, often near crags or streams. They are among the
earliest migrants, some usually arriving in late March. The
white crescent is clearer and brighter on males than females,
and absent on young birds. Their song is less musical than a
blackbird's (p.27) but their loud scolding calls are similar.
See 'Thrushes' p.70

Skylark

Alauda arvensis
Gael: *Uiseag*

Widespread   **All year**

The skylark's *sustained and musical song*, delivered in high-level, *hovering song flight*, is its most distinctive feature. Crouching on the ground, or on a fence post, it looks decidedly *dumpy* and its *short crest* is not always obvious. A liquid 'chirrup' is often given as it flies, when wing beats alternate with closed-wing glides, resulting in an undulating flight pattern. Skylarks are the most widely distributed British species, breeding on farmland, rough grassland, heather moors and hilltops, from sea level to over 1000m. In winter they desert high ground and move nearer the coast.

Meadow Pipit

Anthus pratensis
Gael: *Snàthtag*

Widespread
Mainly March–September

The meadow pipit occupies many of the same habitats as the skylark and is also *streaky-brown* with *white outer tail feathers*, but it is smaller and slimmer and has a very different voice. Its call is a thin 'tseep' and its song, given during its flutter-up and parachute-down song flight, is a not very musical series of thin piping phrases, ending in a trill. Although mainly a bird of open ground it also breeds in young conifer plantations. Many meadow pipits move south in winter, but some remain on low ground.

173

CHATS

Wheatear (p.174), whinchat (p.175) & stonechat (p.175)
- are rather robin-like birds, which stand or perch fairly upright
- frequently flick their tails
- feed mainly on the ground, on insects
- give various 'chak' or 'tik' calls
- usually sing from a prominent perch or in flight
- favour rough open ground with gorse or other bushes (stonechat & whinchat), or stony heath and moorland (wheatear)
- show white/whitish on rump &/or tail in flight; amount and position is important for identification
- sexes differ markedly in wheatear and stonechat, but not in whinchat
- head patterns are distinctive, especially in males

Wheatear

Oenanthe oenanthe
Gael: Brù-Gheal

Widespread March–September
This is *the only small ground-feeding bird with a conspicuous white rump*, which is obvious as the birds fly or hop around, frequently flicking their wings and tails. The wheatear's song is an almost lark-like warble, but much shorter and at times rather creaky and wheezy. This is the most widespread of the chats, occurring on all the islands and in all suitable areas of the mainland. A hole-nester, it frequents open ground with short vegetation and scattered stones or rocks. It is one of the earliest summer visitors to arrive, and also occurs as a passage migrant in spring and autumn.

MALE

FEMALE

MALE

Whinchat

Saxicola rubetra
Gael: Gocan Conaisg

Widespread   May–August

Both male and female have conspicuous *stripes above the eye and at the side of the throat*, white in males and buffish in females. *White patches on the inner wing and the sides of the tail* show in flight.

Whinchats sing from a prominent perch or in short display flight; their short warble starts harshly but becomes sweeter. They breed in areas of rough ground with scattered bushes and in young plantations, which they abandon as the trees close up. They are more widely distributed than stonechats, especially inland in the eastern half of the country.

FEMALE

Stonechat

Saxicola torquata
Gael: Clacharan

Widespread   All year

The male's black 'busby' and white half-collar are distinctive. The female's head is browner and lacks the whinchat's light eye stripe. In flight both sexes can be distinguished from whinchats by the *absence of white on their tails*. The stonechat's song is a rapidly repeated double note, rising and falling. This species is most abundant in the west, where it is resident. In the east and inland it is scarce and local, and generally moves away in winter. It is usually found on rough ground with gorse.

MALE

FEMALE

175

Linnet

Carduelis cannabina
Gael: *Gealbhonn-Lin*

MALE SUMMER

Widespread   All year
Male linnets lose much
of the *red on crown and
breast* in winter, which makes them less easy to identify.
The *whitish patch on the wing* helps in distinguishing females
and young from the larger yellowhammer (p.74) and reed
bunting (p.151), which have *white outer tail feathers* too.
Linnets usually give their slightly nasal twittering song
from an exposed perch, and have a 'tsooeet' call.
They are most abundant south of the highlands
and up the east coast to the Dornoch
Firth, on rough ground with gorse bushes
near farmland, occasionally in hedges and young
conifer plantations.

FEMALE

Twite

Carduelis flavirostris
Gael: *Riabhag Mhonaidh*

MALE SUMMER

Widespread   All year
This species *breeds on
hillsides   and   coastal
moorlands,* mainly in the northwest highlands and the
islands, more locally from the central highlands southwards.
Confusion with the rather similar linnet is
most likely to occur in winter, when flocks of
twites forage for the seeds of
farmland weeds in low-ground areas
where they do not breed; *twites show
less white on the sides of the tail.*

See 'Finches' p.29

FEMALE

THE COAST

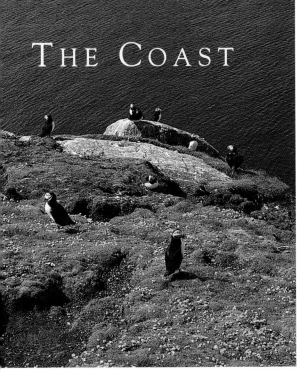

At puffin colonies the cliff-top turf is riddled with burrows, outside which the occupants stand around in sociable groups

Most of the birds which live along the coast feed either at sea or in the intertidal zone on the shore. Rocky coasts provide a variety of nesting sites but limited feeding opportunities, other than offshore, whereas 'soft shores' – especially muddy estuaries – offer good feeding grounds but few suitable places for nesting. Although much of Scotland's coast is rocky, there are also many sandy beaches and estuaries with extensive stretches of intertidal mud. Each type of coastline has its typical bird inhabitants, as do the seas around the islands and in the more sheltered bays and estuary mouths.

The densest seabird colonies are in the north, where kittiwakes and guillemots pack close together on the shelved cliffs

ROCKY COASTS

In summer rocky coasts offer some of Scotland's most thrilling birdwatching, as blizzards of seabirds swirl to and fro near the cliffs. The best place to watch from is where you can sit on the cliff top and look across a narrow inlet at the birds nesting on the opposite side. The numbers and species breeding on a cliff depend upon its formation. Where the rock forms a series of shelves these are usually densely packed with guillemots and kittiwakes; such cliffs are commonest in the north. Cliff faces with smaller ledges and niches accommodate the spaced out nests of fulmar, razorbill and shag, but fewer guillemots and kittiwakes, while gannets make use of both types of situation – provided they are on islands. Great and lesser black-backed and herring gulls often nest along a clifftop, and cliff-girt islands with a turfy hinterland may have colonies of burrow-nesting puffins, manx shearwaters and storm petrels. All seabirds leave the breeding cliffs for at least part of the winter, but some are absent for only a few weeks. Cliffs also provide breeding sites for several species which are not strictly seabirds, among them white-tailed eagle, chough (p.66) and rock dove.

There are few parts of the Scottish coast without rocky islets, low headlands or sea-washed rocks; the bird-life of such coasts varies with the season. In summer two of the most widespread breeding birds, especially in the north and west, are arctic tern and eider duck. In winter the species most closely associated with rocky shores are turnstone and purple sandpiper, which can be seen virtually anywhere that there is exposed seaweed-covered rock. Rock pipits are permanent residents on most rocky coasts, where they feed among dead seaweed; cormorants nest on rocky islets; and herons (p.145) often fish in the shallows along the west coast. Boulder beaches attract few birds other than nesting black guillemots.

SOFT SHORES

Sand and shingle beaches support few insects and small animals and are only poor feeding grounds for birds, so birds often use them simply as resting areas. Common, sandwich and little terns do, however, nest on beaches or among dunes, where

Tide-washed, seaweedy rocks are the main feeding ground of the turnstones and purple sandpipers which winter in Scotland

they may be in company with black-headed gulls. Shelducks and eiders also breed in the dunes, and ringed plovers and oystercatchers on both sand and shingle beaches. In late summer large parties of terns and gulls can often be seen resting on the sand, while during migration periods waders of several species may spend time there. Late summer and spring are the best times for seeing the most typical bird of the sandy beach, the sanderling, which scurries along at the water's edge. And those brave enough to tackle a mid-winter beach walk may be lucky enough to see snow buntings working their way along the wrack at the high tide line. The 'machair' which backs many Hebridean beaches is an especially important habitat for breeding waders. An intricate mosaic of shallow pools, flower-rich grassland and cropped ground, it supports large numbers of nesting ringed plovers, dunlins, redshanks and snipe (p.146).

Muddy estuarine shores are of great importance to birds and special interest to birdwatchers from autumn through to spring. Most such shores are enriched with silt deposited by the rivers flowing into the estuary, and are exposed to a mixture of fresh and salt water. This combination of environmental factors suits small invertebrates such as lugworms and snails, which are present in large enough numbers to sustain big winter populations of waders. For rewarding birdwatching at an estuary it is essential to take account of the tide; when it is fully out the birds are often so far away that they cannot be identified, and when it is fully in many of them fly off to roost on fields or

Not many birds nest on sandy beaches, and those that do expose their eggs to risk from trampling feet, as they can easily be overlooked

headlands. The best time to go is shortly before or after high tide, when the birds are either being forced to come closer as the tide rises or are following it out as it slowly falls. It is best, too, to have the light behind you (which generally means looking from the southern or western shore) and it is easiest to start at a relatively small-scale site, such as the Ythan or Eden, where there is not such an expanse of mud to scan and array of birds to try and sort out.

Knot and dunlin are often the most abundant of the wader species, and good numbers of redshank, curlew (p.62), oystercatcher and ringed plover are usually also present, with smaller numbers of lapwing (p.63), golden plover and bar-tailed

Waders on the mudflats are in constant motion – rising, wheeling and re-alighting – before the incoming tide

godwit. On some estuaries as many as 20 different waders occur annually, but many of these appear only as single birds. Estuaries are also good feeding and roosting grounds for wildfowl. Most attract large numbers of wintering ducks such as wigeon and mallard (p.126); the Solway has barnacle geese (p.58) and pintail; and the Montrose Basin is used by moulting mute swans (p.123) and vast flocks of roosting geese. Shelduck feed over the mudflats for much of the year.

THE SEA

Many of Scotland's seabirds can be watched during ferry trips between islands. Herring gulls regularly accompany boats, hoping to scavenge garbage thrown overboard, and fulmars often follow the wake. Guillemots, razorbills and puffins bob on the surface until the bow is almost upon them before vanishing under the waves. Gannets and kittiwakes gather over fish shoals, circulating vertically as they alternately dive and rise again. Wherever there are such gatherings skuas may also be present, ready to harry any successful fisher. All of these can frequently be seen in summer near the Hebrides and Northern Isles. More difficult to spot, but well worth looking out for, are manx shearwater and storm petrel.

In autumn and winter inshore waters are frequented by divers and ducks, and in a few places also by grebes. Where there are good supplies of mussels and crabs seaducks such as eider, common and velvet scoter and long-tailed duck congregate, for example in the Moray Firth and around the mouth of the Tay, while shoals of sprats in the Beauly Firth attract goosanders and red-breasted

Inter-island ferries offer good opportunities for watching seabirds at sea

mergansers, and goldeneye and scaup gather near sewage outfalls. Both cormorants and shags are present throughout the year, the former often in estuary mouths while the latter prefers open sea coasts. Red-throated divers (p.134) move southwards off the east coast in early autumn, when there are often also spectacular south-bound movements of seabirds, most readily visible from prominent points such as Rattray Head, Fife Ness and St Abbs Head.

ADULT

Gannet

Sula bassana
Gael: Sùlaire

Widespread   **Almost all year**

Goose-sized and *conspicuously white and black*, adult gannets
are more distinctive than immature birds, which have
variable amounts of brown on back and wings. They nest on
islands, mainly in the far north and west, but also in large
colonies on the Bass Rock and Ailsa Craig. Even in areas
where they do not breed gannets can often be seen
offshore, flying in 'strings' or circling and
plunge diving from a height with folded wings.
The nesting colonies are occupied from
March to August, and gannets are scarce in Scottish
waters only between December and February.

IMMATURE

White-tailed Eagle

Haliaeetus albicilla
Gael: Iolair-Mhara

ADULT

Local   **All year**

This huge bird has been described as looking like a flying
door! It has a *wedge-shaped tail* and unfeathered lower legs,
whereas the golden eagle's (p.162) tail is square ended and
its legs are feathered. White-tailed eagles were
exterminated by persecution early this century but were
recently reintroduced on Rum. Several pairs now breed on
the west coast north of the Clyde, and wandering birds are
occasionally seen in other areas, sometimes inland.

Shag

ADULT

Phalacrocorax aristotelis
Gael: Sgarbh an Sgùmain

Widespread   **All year**

Shags are more widespread and more maritime than cormorants, and breed on all exposed rocky coasts, usually nesting on low ledges or in caves. Their *plumage has a green/bronze gloss*, and their *short crests* show best when blowing in the wind. Shags at the nest often hiss with wide open bills, exposing their *bright yellow mouths*. The dark brown young birds lack the whitish bellies of young cormorants. When swimming shags hold their heads horizontal, not uptilted. Like cormorants they often stand with wings spread, but seldom perch on anything other than rocks.

JUVENILE

Cormorant

ADULT SUMMER

Phalacrocorax carbo
Gael: Sgarbh

Widespread   **All year**

Its *white lower face* distinguishes this species from the smaller shag, and it frequents sheltered coastal waters rather than sites exposed to rough seas. Cormorants swim low in the water with head and bill tilted upwards; they dip their heads below the surface and dive frequently. They readily perch on buoys and posts, sitting upright and often with wings 'hung out to dry'. Most nest on rocky islands, but in winter some move inland to lochs and rivers, where they regularly perch on trees.

JUVENILE

Great Skua

Stercorarius skua
Gael: *Fàsgadan*

Local

ADULT

At colonies April–August

Their *heavy build*, darker brown colour, *conspicuous white wing patches* and short tails distinguish great skuas from young gulls and arctic skuas. They harass for their fish catch, and sometimes kill, birds up to gannet size. At their moorland colonies, most of which are in the Northern Isles, the breeding birds dive bomb intruders; 'clubs' of immatures gather near freshwater pools. Great skuas often stand with raised wings, and have a gruff 'uk-uk-uk' call. They move south off mainland coasts in August–September.

ADULT

Arctic Skua

ADULT PALE

Stercorarius parasiticus
Gael: *Fàsgadair*

ADULT DARK

Local
At colonies April–August

A dark-backed 'gull' with *extended central tail feathers*, this piratical bird flies fast and gracefully, twisting and turning as it chases kittiwakes and terns and dives to catch the fish they drop. Near its moorland nest it pretends to have a broken wing to distract attention from egg or chick, and dive bombs intruders aggressively. Its commonest calls are a sharp 'ya-wor' when alarmed, and a wailing 'eee-air'. Arctic skuas breed mainly in the Northern Isles; in autumn they pass down both east and west coasts on migration. This species occurs with white, dark and intermediate underparts.

ADULT PALE

187

GULL-TYPE BIRDS OF ROCKY COASTS

In decreasing order of size:

The three large gulls – great black-backed (p.190), herring (p.191) and lesser black-backed (p.190)

- stand tall and walk readily
- fly with strong wing-beats, usually showing a bend at front of wing
- adults have yellow bills with a red spot
- adults have very dark wing tips with white spots
- adults are distinguished by colour of back and legs
- young birds are mottled grey brown and difficult to distinguish

Fulmars (p.189)

- characteristically squat with legs hidden
- glide with straight wings
- have a 'tube' on upper surface of bill and no red spot
- are uniformly coloured on back, wings and tail
- have no black on wing tips

Kittiwakes (p.191)

- have short black legs
- have yellow-green bills, red inside the mouth
- have black wing tips without white spots

Arctic terns (p.196)

- have slim bodies, narrow wings and deeply forked tails
- have short red legs, black caps and blood red bills
- have no black on wing tips

Juvenile herring gull

Fulmar

Fulmarus glacialis
Gael: *Fulmair*

Widespread   **November–September**

Fulmars *glide on stiffly-straight wings*, using their large webbed feet as rudders, and often hang motionless on up-currents along the cliff face. At sea they fly low over the water, tilting first one way and then the other, and flap only occasionally. They differ from gulls in having 'tubenoses', *heavy-looking beaks with an obvious nasal tube, and no black on the wings*. They cackle and head wave at the nest but away from the cliffs are generally silent. Both adults and chicks sit back 'on their haunches' and *never stand up straight-legged like a gull*.

Fulmars feed on the ocean surface, floating buoyantly as they forage for small marine creatures and offal discarded by fishing boats. They nest separately, usually on a ledge overhanging the sea, on cliffs all round the coast and also in a few places inland (eg. Salisbury Crags in Edinburgh). Fulmars are entirely absent from their breeding cliffs only from mid-September until November.

See 'Gulls' p.188

Great Black-backed Gu

Larus marinus
Gael: Farspach

ADULT

Widespread   **All year**

This is the *largest and darkest gull. Its legs are pale pink.* It breeds most abundantly in the north and west; few nest on the east coast south of Aberdeen. Great black-backs prefer breeding sites not easily accessible to man and rarely nest in large colonies. They are predatory and rather vicious, killing adult and young birds, and often feed on carrion. In winter they occur all round the coast, with the biggest gatherings around fishing ports or near coastal rubbish tips; they are less often seen inland than the herring gull (p.191).

Lesser black-backed Gu

Larus fuscus
Gael: Farspach Bheag

ADULT

Widespread   **Late February–September**

Lesser black-backs have *yellow legs.* Most breed along rocky coasts in the Northern and Western Isles and on the northwest mainland, though there are colonies on the Isle of May and inland at Flanders Moss. They nest on rocky headlands, rough ground on small islands, and moorland rather than clifftop positions. Most move away from Scotland in winter; when returning from the south in February and March lesser black-backs often visit farmland.   *See 'Gulls' p.188*

Herring Gull

Larus argentatus
Gael: Glas-Fhaoileag

ADULT

Widespread   **All year**

The *palest, commonest and most widespread of the big gulls*, this species breeds on all rocky coasts and occasionally inland and on rooftops in seaside towns. Breeding pairs are noisy and often aggressive and may dive-bomb intruders. This behaviour, together with their loud and strident early morning 'kyow' calls, makes them decidedly unpopular with human neighbours! In winter the largest concentrations are around east coast fishing ports and near garbage tips in the central lowlands. Herring gulls scavenge almost any edible refuse and regularly follow boats at sea.

ADULT

Kittiwake

ADULT

Rissa tridactyla
Gael: Seagair

Widespread   **March–September**

Much the *smallest of the cliff-nesting gulls*, the kittiwake has *short black legs*, a *yellow-green bill*, and a *red mouth* which is easily seen when it calls 'kitti-wak'. It is a maritime species, coming to land only at its colonies, where its mud and seaweed nest is attached to a tiny ledge on the cliff face. Kittiwakes plunge dive and often gather in flocks over shoals of small fish well offshore. They are most abundant in the Northern Isles and on suitable north and east coast cliffs, scarcer in the west.   See 'Gulls' p.188

JUVENILE

Guillemot

Uria aalge
Gael: Eun Dubh an Sgadain

SUMMER BRIDLED

SUMMER

Widespread

**At colonies
March–early July**

Guillemots have *slender pointed bills;* some have a white 'bridle' round the eye. They are most abundant in the north and east; at their colonies they keep up a noisy trumpeting 'aargh' and form rafts on the sea nearby. They pack close together on their breeding ledges, facing towards the cliff, which helps to prevent egg or chick from falling into the sea; the young leave before they are able to fly. Guillemots surface dive for fish, swim fairly high in the water, and winter at sea; oiled birds are found on beaches most winters.

WINTER

Razorbill

SUMMER

Alca torda
Gael: Coltraiche

Widespread

**At colonies
March–July**

Razorbills have a *conspicuous white vertical line near the tip of the deep bill* and are blacker than guillemots. They are much less numerous than guillemots and, like them, are most abundant in the north, though breeding on suitable cliffs right round the coast. They usually nest singly, in cliff crevices or among boulders, and can often be seen offshore in family parties in late summer. They winter at sea. When taking off from the water they patter along the surface.

WINTER

Puffin

SUMMER

Fratercula arctica
Gael: *Fachach*

Widespread   At colonies April–July

Their *brightly coloured bills and white
cheeks* make puffins unmistakable in
summer, as they stand around near
their nesting burrows, which are
usually on steep grassy slopes. They
are commonest on islands in the north and
rather scarce south of Caithness, except on the
Isle of May. Puffins gather in rafts on the sea
near their colonies, looking even dumpier on
the water than ashore, and fly to and fro with
rapidly whirring wings. Their bills lose much of
their colour in winter, which is spent far at sea.

WINTER

SUMMER

Black Guillemot

Cepphus grylle
Gael: *Gearradh-Breac*

Widespread   All year

This is by far the *smallest* and
scarcest *of the auks*, and is
readily identified by its *white
wing patch*, conspicuous at rest and in flight. Black
guillemots, which stay inshore all year, are thinly scattered
on rocky coasts and most abundant in the Northern Isles.
They nest solitarily, in rock crevices and on boulder
beaches, but gather to display early in the breeding season.
Their high pitched whistling cry is given with the beak
open, showing the red mouth. In winter they look very
different, with mottled grey-
white back and sides.

WINTER

Storm Petrel

Hydrobates pelagicus
Gael: Luaireag

Local   **At colonies**
                   May–October

This *tiny* seabird is most likely to be seen from a boat, as it *flits low over the waves*, its feet often pattering the surface of the water. Storm petrels breed on islands in the north and west, and change over at their nesting burrows at night. They have a musty smell and a peculiar purring call ending in a hiccup; both can be experienced on Mousa, Shetland, where the birds nest in the dry stone walls of the ancient broch. They winter far at sea.

Manx Shearwater

Puffinus puffinus
Gael: Sgrab

Local   **At colonies**
                   March–August

Manx shearwaters are most likely to be seen at sea, where they *look alternately black and white as they glide low over the waves*, tipping first one way and then the other and only rarely flapping. They breed in a large colony on Rum and a few smaller ones on islands in the west and north, and on summer evenings gather in rafts off these sites. As they fly to their burrows at night they give weird screams and cackles. In autumn parties can be seen all round the coast, as they leave Scottish waters to winter far at sea.

TERNS & SMALLER GULLS

Terns – arctic (p.196), common (p.196), sandwich (p.197) & little (p.197)
- are migratory & mainly coastal
- have slim bodies, short legs, narrow wings and deeply forked tails
- in summer have black caps contrasting with pale grey and white plumage
- hover and plunge-dive for fish in shallow water
- nest on the ground, usually colonially, and are sensitive to disturbance
- have shrill, rather harsh 'kee-ya' and /or 'kier-ik' calls
- rarely walk away from the nesting area

Arctic tern

The smaller gulls – common (p.142) & black-headed (p.143)
- are resident and widespread inland as well as on the coast
- have plumpish bodies, medium long legs, and broad rounded tails
- feed on the ground or by hawking for insects
- nest on the ground, in marsh vegetation, and occasionally on trees
- have squealing or scolding 'kya' or 'ker' calls
- walk readily and often follow the plough

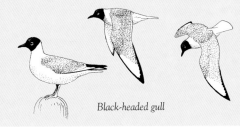

Black-headed gull

ADULT SUMMER

Arctic Tern

Sterna paradisaea
Gael: Steàrnal

Widespread   **Late April–August**

This tern's *bill is blood red right to the tip*, whereas a common tern's has a black tip. When seen overhead the *points of its wings look semi-transparent*. It is often difficult to be certain of sight identification, but location is a useful guide to the most likely species. Arctic terns are by far the most abundant terns in the Northern Isles, more thinly scattered in the west and relatively scarce on the east coast. They nest colonially on rocky islands and coastal moorland, and are aggressive, often dive bombing intruders.

ADULT
AUTUMN

ADULT SUMMER

Common Tern

Sterna hirundo
Gael: Steàrnag

Widespread   **Late April–August**

A black tip to its *orange-red bill*, longer legs and a shorter tail are the most obvious differences between this species and the arctic tern. *Only a small patch on the wing looks semi-transparent* from below, not the whole tip. Common terns nest on sand and shingle beaches, and occasionally inland on river shingles. They breed in almost every county and are the commonest terns in the south west and in the east from Aberdeen southwards. Their colonies seldom exceed 100 nests, whereas more than 1,000 pairs of arctic terns sometimes occupy a site.

See 'Terns' p.195

ADULT SUMMER

Sandwich Tern

ADULT SUMMER

Sterna sandvicensis
Gael: Steàrnag Mhòr

Local   **April–August**

This is the *largest and most gull-like of the terns* and is easily distinguished by its *untidy black crest* and mainly *black bill*. It is much scarcer and more local than common and arctic terns, and breeds at only a few sites. The most regularly occupied colonies are in Orkney, East Ross, Grampian – at Sands of Forvie and the Loch of Strathbeg – and on the island of Inchmickery in the Forth. Sandwich terns, like the other species, gather on beaches in July and August before migrating.

ADULT SUMMER

Little Tern

ADULT SUMMER

Sterna albifrons
Gael: Steàirdean

Local   **May–July**

Its *small size, mainly yellow bill* and *white forehead* readily identify the little tern, which is much scarcer than the other species. Its colonies are widely but very thinly scattered, usually on sandy beaches, and seldom comprise more than a few pairs. Because they choose beaches attractive to man, and frequently lay close to high tide mark, little terns often fail to breed successfully unless given special care and protection on wardened reserves.

See 'Terns' p.195

JUVENILE

Shelduck

Tadorna tadorna
Gael: Cràdh-Ghèadh
Widespread   **All year**

This *large and distinctively coloured* duck *looks black and white at a distance*. It is more often seen feeding over wet mud or sand, with a scything motion of the head, or standing around near its nesting burrow, than on the water. Shelduck breed solitarily beside sandy coasts and estuaries, with a very few nesting inland; they are territorial and have frequent noisy encounters with neighbouring pairs. In August–October most leave the breeding areas to moult, many going to the inner Forth. Young shelduck are brownish above and white below.

ADULT

JUVENILE

Pintail

Anas acuta
Gael: Lach Stiùireach

MALE

Local   **Mainly September–March**

Flocks of pintail occur regularly only on the Solway, Moray, Cromarty and Dornoch Firths, but single birds or small parties occasionally appear elsewhere on the coast and on lochs inland. The *drake's white neck and long tail* make it easy to recognise. Female pintail have longer necks than mallard (p.126) or wigeon, with which they are most likely to be confused; they are slimmer than mallard and have grey bills and more pointed tails. Pintail fly fast, with quicker wingbeats than mallard, and usually feed by up-ending.   *See 'Ducks' p.124-5*

FEMALE

Wigeon

Anas penelope
Gael: Glas-Lach

MALE

Widespread   **All year**

Male wigeon can be distinguished at all seasons in flight by their *conspicuous white forewing*; females, and males not in their distinctive breeding plumage, are a much *redder brown than the larger mallard* (p.126) and have proportionately *smaller, blue bills*. Wigeon rise straight from the water and their flight is fast and direct. Drakes give a whistling, musical 'whee-oo' and ducks a purring growl.

Wigeon breed in small numbers near pools and lochs on moorland, mainly in the central and northern highlands but also locally in southern Scotland and a few of the islands. In winter many more are present and they are much more widely distributed, on lochs and estuaries and also in some areas on grassland along river banks. They feed by up-ending but also regularly graze on

MALE

short grass and on eelgrass growing on estuarine mud, and when on freshwater often scavenge plant material brought up by coots and swans. Wigeon are very gregarious in winter and on east coast estuaries are frequently present in thousands, greatly outnumbering mallard and teal (p.127), the other dabbling ducks likely to share such sites.

See 'Ducks' p.124-5

FEMALE

199

Eider

Somateria mollissima
Gael: Colc

MALE

Widespread   **All year**

Moulting and young eider drakes are often patchy dark brown and white, but can always be identified by the *long even slope from bill to crown*. The *female is the only large uniformly mottled brown sea duck*. Eiders both surface dive and up-end in the shallows and feed largely on mussels, leaving heaps of crushed shells in their droppings. They are very gregarious, often gathering in large rafts offshore outside the breeding season; their take-off from the water is laboured, and they usually fly low and in single file.

Eiders breed on all open coasts, most often in rocky areas but also locally among dunes. Courting males croon 'ah-ooo' in chorus and perform an elaborate head-waving display. Incubating females are frequently very tame, but are so well camouflaged that they are not easy to see. A creche system is often operated, with a few females looking after several broods of sooty brown chicks, which sometimes have to walk a long way, or descend cliffs up to 100m high, to reach the sea. In July–September moulting flocks gather off the coast, the largest at the mouth of the Tay.

FEMALE

See 'Ducks' p.124-5

Common Scoter

Melanitta nigra
Gael: Tunnag Dhubh

MALE

Widespread   **All year**

This seaduck occurs most regularly off the east coast, in both summer and winter, and is *often visible only as a tight raft of all-dark birds well offshore.* All members of a flock frequently dive at the same time, so that the whole raft vanishes from sight. Moulting flocks are present in July–August off Dornoch, north of Aberdeen, near Montrose, and in the Forth off Gullane; the main wintering flocks are in the Moray Firth, St Andrews Bay, and the Forth.

FEMALE

Velvet Scoter

Melanitta fusca
Gael: Lach Dhubh

Widespread   **Mainly July–May**

Velvet scoters often accompany common scoters, from which they can readily be distinguished in flight, or when flapping their wings, by their *white wing bar.* On the sea at a distance it is difficult to identify them with certainty.

MALE

See 'Ducks' p.124-5

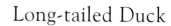

Long-tailed Duck

Clangula hyemalis
Gael: Lach Bhinn

MALE WINTER

Widespread **October–April**

Long-tailed ducks are active and agile surface divers, whose *small size, round head and small bill*, and *piebald pattern* help to distinguish them from other seaducks. They winter in large numbers in the Moray and Dornoch Firths, and the Northern Isles, and in smaller groups elsewhere, especially off the east coast. In spring the males give a musical yodelling call, often in chorus. Individuals occasionally turn up inland.

FEM

Scaup

Aythya marila
Gael: Lach Mhara

MALE

Rather local **October–March**

Scaup are usually seen at the coast, where they winter in large flocks at a few sheltered sites: the most regular of these are the Forth near Leven, the Solway between Carsethorne and Southerness, Loch Ryan, the Clyde, Loch Indaal on Islay, and the Dornoch Firth. Small numbers occasionally appear on inland freshwaters, where *males can be distinguished from drake tufted ducks (p.130) by their grey backs*. *Females* can be confused with female tufted ducks, which sometimes also *have a white band around the base of the bill*, though this is usually broader and more conspicuous in the scaup.

See 'Ducks' p.124-5 FEMAL

WADERS

Waders
- vary widely in size, and length of bill and legs:
  large: curlew (p.62), oystercatcher (p.204), bar-tailed
  & black-tailed godwit (p.209), whimbrel (p171),
  woodcock (p.87)
  medium: lapwing (p.63), golden (p.170) & grey (p.206)
  plover, redshank (p.208), greenshank (p.171), knot
  (p.206), snipe (p.146)
  small: ringed plover (p.205), dunlin (p.207), sanderling
  (p.203), purple (p.210) & common (p.147)
  sandpiper, turnstone (p.210), dotterel (p.170)

Useful clues to identification are:
- habitat & feeding method
- length & colour of legs & bill
- position of any white patches/bars on wings, rump &/or tail
In autumn many coastal waders are young birds, which
  are drabber and often browner than adults
Adults in spring and early autumn often show some of
  their breeding season colours

Sanderling

Calidris alba
Gael: Luatharan Glas

Widespread　　　　　　　**Mainly August–May**

This *plump pale little wader* is
usually seen *running like a
clockwork toy along the very edge of the retreating waves* on
sandy beaches, stopping occasionally to peck in the wet sand
with head low and tail high. Sanderling often call with a
liquid 'twik-twik' as they fly low over the sea, and their
wings appear to flicker due to their grey, black and
white pattern. Some sanderling are present all
year, but numbers in most areas are highest in
August–September and May, when flocks are
passing through on migration.

Oystercatcher

Haematopus ostralegus
Gael: Gille-Brìghde

ADULT

Widespread   **All year**

Both in flight and on the ground the oystercatcher's *black and white plumage and long orange-red bill* make it difficult to confuse with any other species. In summer it is one of the most widely distributed of the waders, breeding along the coast and also inland in fields, near rivers and on loch shores. Birds move inland from February onwards, when their loud 'kleep' contact calls can often be heard as they fly overhead at night. Oystercatchers are very noisy, and on territory keep up prolonged piping duets and repeated 'pik-pik-pik' calls well into the night. They start to desert inland areas for the coast in July.

ADULT

In winter oystercatchers are scattered along all types of low coast, but the main concentrations are in the larger estuaries. They feed by stabbing into mud or soil, or poking around in rock pools and among seaweed, and can open shellfish with their strong bills. When feeding they are usually well spaced out, but at high tide they gather in tight flocks to rest on islets, sand bars and occasionally fields.

See 'Waders' p.203

Ringed Plover

Charadrius hiaticula
Gael: Bòdhag

ADULT

Widespread   **All year**

The ringed plover's *pied head and breast and white neck band* distinguish it from other common small waders. It is *plump and compact*, and its *behaviour* helps with its identification: it looks *alert and busy* as it moves in short runs, stopping to tilt forward to pick up food, then running on again.

Most ringed plovers nest on or near sand or shingle beaches, but some move far inland to breed on loch shores and river shingles; especially large numbers breed on the 'machair' of the Outer Hebrides. When intruders approach a nest the birds try to lure them away, by feigning a broken wing or fanning their tails and squealing like a rodent. They keep up a continuous plaintive 'queep'ing while an intruder is present. In winter the majority of ringed plovers are in estuaries, but many remain on Hebridean beaches and there are some on most sandy or muddy coasts. When feeding this species behaves differently from the two other similarly-sized common waders, with individuals staying well scattered over the mud, and neither forming tight flocks like dunlins (p.207), nor running along the water's edge like sanderlings (p.203).

See 'Waders' p.203

ADULT

Knot

Calidris canutus
Gael: Luatharan Gainmhich

Widespread   July–March

Knots are among the most
gregarious of the waders and usually
occur in tightly-knit, and
often large, *flocks*. They feed

ADULT WINTER

close together, keeping up a chorus of muttered 'nut's, and
fly in aerobatic clouds, looking alternately light and dark as
they wheel and turn. They are noticeably *dumpy but
otherwise rather nondescript*, though in early autumn and
late spring there may be traces of their russet
breeding plumage on back and underparts.
The main winter concentrations of knot are on
the Forth and Solway, but some are present on
most estuaries, and they also occur in small
numbers on other types of shore.

ADULT

Grey Plover

Pluvialis squatarola
Gael: Trilleachan

Local   August–April

This species can most easily be distinguished
from the much commoner golden plover (p.170)
in flight, when its *whitish rump and wing bar and
black 'armpits'* can be seen. On the ground it
might be confused with a knot, but is taller, has a
shorter bill, and occurs in small loose
parties, not tight flocks. The grey
plover is a rather scarce winter visitor,
which frequents muddy estuaries and
occurs most regularly on the Solway, Forth
and Eden.

See 'Waders' p.203

Dunlin

Calidris alpina
Gael: *Pollaran*

ADULT SUMMER

Widespread   **All year**

Dunlins winter on the coast and breed on moorland. When feeding over estuaries and beaches they move forward together in flocks, with heads down as they probe and pick in the mud; their *shoulder-hunched attitude* and fine *slightly down-curved bills* help in identification. Flocks frequently perform aerial evolutions, 'flowing' to and fro in a tight cloud, sometimes close to a flock of much bulkier knots (p.206). Dunlins in winter plumage lack conspicuous

WINTER

markings but in flight the *narrow whitish wing bars and white sides to the rump* usually show up well. Their summer plumage, with the *black belly patch*, is distinctive.

From August to March dunlins are widely scattered around the coast, with large gatherings on the Inner Clyde, the Solway and many east coast estuaries. Their most important breeding grounds are in the Northern Isles, Caithness and Sutherland, and the Uists, but they are thinly scattered in suitable areas throughout Scotland. They nest solitarily, at altitudes ranging from sea level in the islands to about 1000m in the Cairngorms. Nesting pairs attract attention with their shrill 'referee's whistle' calls, or the male's trilling song given in display flight.

See 'Waders' p.203

WINTER

207

Redshank

Tringa totanus
Gael: *Maor-Cladaich*

SUMMER

Widespread   **All year**

Redshanks breed on low moorland, damp pasture and marshy ground, and winter on the coast. Their *long bright orange-red legs, conspicuous white wing bar and rump, and noisy calls* make them easy to identify. They make a variety of yelping sounds, the commonest of which is 'tyu-yu-yu', with emphasis on the first syllable. They often stand on fence posts, bobbing up and down when alarmed, and fly fast, tilting from side to side, and frequently holding their wings upraised after landing.

Redshanks are widely, but rather thinly, distributed on their breeding grounds from April to August; their numbers have decreased in recent years as farmland drainage has reduced the amount of suitable habitat. When on breeding territory they give yodelling and repeated 'yip-yip' calls. In winter there are redshanks on all except cliff coasts, with by far the greatest numbers concentrated in the larger muddy estuaries. Feeding birds are usually well scattered as they walk about probing in the mud. In mixed flocks of waders redshanks look smaller than godwits, slimmer and longer-legged than golden (p.170) and grey (p.206) plovers and knot (p.206), and much larger than dunlin (p.207).

See 'Waders' p.203

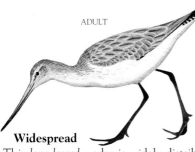

ADULT

Bar-tailed Godwit

Limosa lapponica
Gael: *Roid-Ghuilbneach*

Widespread   **August–March**

This *long-legged* wader is widely distributed on muddy and sandy shores, with large numbers wintering on the major east coast estuaries and the Solway; elsewhere parties tend to be quite small. Bar-tailed godwits feed by probing with their slightly upturned bills as they walk along the tideline, or in shallow water with head and neck submerged. They are intermediate in size between redshank and curlew (p.62), and are most likely to be confused in flight with curlew or greenshank (p.171), which also have *white rumps* and *no wing bars*. Flocks of bar-tailed godwits, which are usually silent, often perform aerial manoeuvres.

Black-tailed Godwit

Limosa limosa
Gael: *Cearra-Ghob*

Local   **August–April**

This species is much scarcer than the bar-tailed godwit, from which it can be distinguished by its *conspicuous white wing bars and straight bill*. It winters in small numbers on the larger estuaries, but is much more widely scattered on muddy shores during migration periods, in August–September and March–April. In spring many black-tailed godwits show signs of their chestnut breeding plumage, especially on head and neck.

See 'Waders' p.203

Purple Sandpiper

Calidris maritima
Gael: Cam-Ghlas

Widespread
Mainly August–April

This *dumpy little* wader is closely associated with low rocky shores which have seaweedy areas exposed at low tide. Its *dark back and upper breast* help it to blend into its background, but it is so tame that its *yellow legs* can generally be seen without difficulty. Purple sandpipers are usually in small parties, often with turnstones, and *twitter in chorus*, sounding rather like swallows. They fly low over the water and land with a sudden flutter; in flight the white sides to the tail contrast with the very dark back.

Turnstone

Arenaria interpres
Gael: Trilleachan Beag

ADULT WINTER

Widespread

Mainly August–early May

The turnstone's *pied flight pattern* is distinctive among small waders, and easily separates it from the purple sandpiper, which also frequents low rocky shores. Turnstones wintering here have predominantly grey-brown backs but these are often speckled with chestnut, especially in autumn and spring. They are gregarious and quarrelsome, and give frequent staccato 'tuk-a-tuk' calls as they *bustle about turning over stones and seaweed* with their short bills. Turnstones are most abundant in the east and north.

See 'Waders' p.203

ADULT WINTER

Rock Dove

Columba livia
Gael: *Calman Creige*

Widespread   **All year**

Many feral pigeons look like rock doves, from which they are descended. Pure rock doves are now found only on the north and west coasts, breeding in caves on both mainland and islands. They are *smaller than woodpigeons* (p.69) and have *no white on the neck or wing*; in flight their *white rump* and *double black wing bar* are conspicuous. Rock doves seldom perch in trees, preferring to settle on the ground, rocks or cliff ledges.

Rock Pipit

Anthus spinoletta
Gael: *Gabhagan*

Widespread   **All year**

This common small bird of tidal rocky shores spends much of its time searching for insects among rocks and tide-wrack, where it blends into its background. It is *darker than a meadow pipit* (p.173) and has *grey, not white, outer tail feathers*, similar thin 'seep' calls and a less tuneful song consisting largely of 'see' sounds. Rock pipits do not flock, even where there are many present, as is the case on much of the north and west coasts. Unlike meadow pipits, rock pipits remain in the same area all year.

Snow Bunting

Plectrophenax nivalis
Gael: Eun an t-Sneachda

MALE WINTER

Widespread   **Mainly October–March**

Adult snow buntings are *white below at all seasons* and *in flight show a lot of white on wings and tail*, which is why they are sometimes known as 'snowflakes'. A very few pairs breed on the highest mountains, but many more winter here, when flocks rove around, sometimes joining up with other finches on the fields, but also feeding among tidewrack on beaches and marram grass on sand dunes, and near the skiers' car parks at Cairngorm and Glenshee. They call frequently, 'trrip' or 'tyu', and have a dancing flight.

See 'Buntings' p.29

FEMALE SUMMER

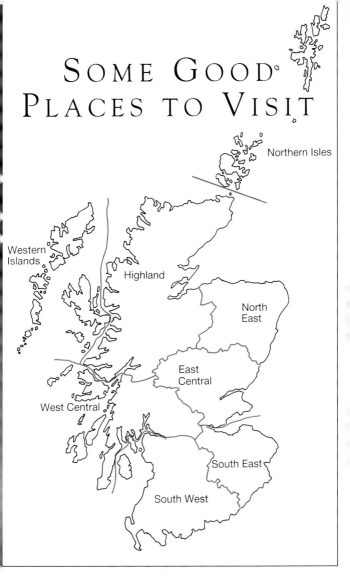

SOME GOOD PLACES TO VISIT

Northern Isles

Western Islands

Highland

North East

East Central

West Central

South East

South West

The sites included in this chapter have been chosen to represent examples of the different habitat types that occur in each of the eight areas shown on the maps; not all habitats occur in all areas. No specific farmland sites are mentioned, but minor roads often provide views over farming country and opportunities to walk alongside fields and hedges. Some of the places listed are single sites, while others are localities where several sites of different types are grouped close together. All the sites included are open to the public all (or nearly all) year, but some have limited opening hours and some an admission charge. Details of many more good birdwatching places are given in *Where to Watch Birds in Scotland*, *A Guide to the Nature Reserves of Scotland*, the RSPB's leaflet *Nature Reserves – information for visitors* and the SWT's *Reserves* leaflet (see Appendix). There are also many town and country parks with good areas of woodland or freshwater, private estates which are open to the public on a regular basis or for special Gardens Open days, and right of way footpaths which go through farmland or upland areas. Details of these can be obtained from local tourist information offices. Many such places offer good opportunities for birdwatching.

THE SITE DESCRIPTIONS & MAPS

The notes on location and access include the name of the nearest town or village and the approach route, and the brief description of each site includes reference to any hides, trails or visitor centres. The general bird interest of each site is described but no attempt is made to list all the species which may be seen. Where the interest is largely seasonal the best time to visit is indicated. The following abbreviations are used:

LNR Local Nature Reserve
NTS National Trust for Scotland
NNR National Nature Reserve
RSPB Royal Society for the Protection of Birds
SWT Scottish Wildlife Trust

One or two of the principal towns in each area are shown on the maps to assist with location. Numbering of the sites is roughly clockwise from the largest town, but the larger islands grouped with a mainland area are treated last in the section.

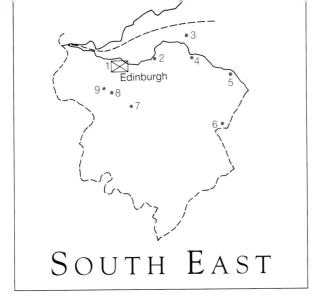

SOUTH EAST

1 Edinburgh Parks

*Holyrood Park & Duddingston Loch: car park & SWT visitor centre (summer only)
near Holyrood Palace and car park beside Duddingston Loch.*

Fulmars breed on Salisbury Crags (one of the best known
inland sites), and skylark, meadow pipit and linnet on
Arthurs Seat. Duddingston Loch has nesting feral greylag
geese, several duck species and grebes. Heron and cormorant
are frequent visitors and there is a varied population of small
birds in the reedbeds, scrub and nearby woodland.

*Hermitage of Braid: access off Braid Road, south of Morningside, or via path from
Blackford Pond. Nature Trail.*

In summer this mixed semi-natural woodland has a good
range of woodland and scrub species, including wood
warbler and both woodpeckers.

Royal Botanic Gardens: Inverleith Row.

Among the woodland and garden birds frequenting the
park is the very local hawfinch.

2 Aberlady Bay LNR, Gullane & Gosford Bays

Access from A198 between Longniddry and Gullane. Car parks at Aberlady Bay (restricted), Gullane Beach and Gosford Bay. Warden at Aberlady Bay. A footpath runs from Aberlady through the reserve to Gullane.

Between them these sites offer a good variety of spring, autumn and winter birdwatching. Aberlady Bay's extensive tidal mud and sand flats support large numbers of dabbling ducks and waders, while flocks of pink-footed geese flight from nearby farmland to roost there. Flocks of finches and occasionally snow buntings often feed over the saltmarsh. Seaducks, grebes and divers can be watched from Gullane Point, and Gosford Bay is especially good for wintering grebes. In summer seabirds can be seen offshore, and in autumn migrating skuas move in to harass the terns and gulls.

3 Bass Rock, North Berwick

In summer there are regular boat trips round the Bass from North Berwick Harbour.

The Bass is home to a large colony of breeding seabirds, including several thousand pairs of gannets and guillemots, and smaller numbers of razorbills and puffins. The packed cliff ledges can be well seen from the boat, which moves slowly round the Rock close inshore. Landing is possible only by prior arrangement (contact Fred Marr, boatman, North Berwick 2838 for details).

4 John Muir Country Park, Dunbar

Signposted off A1087 between A1 and Dunbar. Car parks, ranger service.

The Park's main interest centres on the estuary of the River Tyne, with its inter-tidal mudflats and saltmarshes. In winter these attract large numbers, and a good variety, of wildfowl and waders – including whooper swans, grey plover and bar-tailed godwit – which in turn bring in hunting birds of prey. Most of the seaducks, and often sawbills too, can be watched from the rocky headland at the east of the bay. The Park is too busy in summer to have many breeding ducks or waders but a few eiders, shelduck and ringed plovers do nest, and there are kittiwakes on Dunbar cliffs.

5 St Abbs Head NNR, 2 miles north of Coldingham.

A1107 to Coldingham, then B6438 to car park, visitor centre and coffee shop at Northfield Farm. Ranger service.

April to July is the best time to see the vast numbers of seabirds which breed on this spectacular stretch of cliff coastline; they can be most easily watched from the area around the lighthouse. The cliffs also provide good vantage points for watching offshore movements of seabirds such as skuas in spring and autumn. Tucked down behind the cliffs lies the Mire Loch, surrounded by trees and scrub among which small migrants often shelter when easterly winds are blowing in April–May and August–October. This reserve is owned by the NTS and managed jointly with the SWT.

6 The Hirsel, Coldstream

Signposted from A697 west of Coldstream. Car park, museum, hide.

With extensive policy woodlands, a shallow reed-fringed lake, and a stream running through pasture land, this privately-owned estate supports a wide range of breeding birds. The mixed woodland, some with a dense understorey of rhododendron and other shrubs, provides ideal habitat for warblers, woodpeckers, flycatchers and two very local species, the marsh tit and hawfinch. Most of the common waterbirds breed on the lake, which is over-looked by a hide, and many more come in to roost in autumn and winter. In summer common sandpipers and grey wagtails nest by the stream, which has resident dippers and visiting kingfishers.

7 Gladhouse Reservoir, Howgate

By minor road from A703 or B6372. View from road along north shore.

This large reservoir attracts a variety of wintering duck, and is one of the best places for watching flocks of pink-footed geese flighting in to roost at dusk. When water levels are low in late summer and early autumn migrating waders often visit the shores.

8 Pentland Hills Walk

From Bavelaw, c2^1/$_2$ miles south of Balerno (approached off A70), a signposted footpath runs through the hills of the Pentland Regional Park, via Glencorse Reservoir, to Flotterstone on the A702. Information point at Flotterstone, ranger service.

A variety of upland birds can be seen along the route, among them red grouse and in summer whinchats, golden plovers and ring ouzels, with common sandpipers by the reservoirs.

9 Threipmuir Reservoir & Bavelaw Marsh, Balerno

Approach via A70 and Balerno. Car park at Red Moss, c2 miles south of Balerno.

Much of the open water and marshland can be viewed from the two bridges which cross it; a path runs along the north shore of Threipmuir. Breeding water birds include grebes, feral greylag geese, and a large black-headed gull colony. A variety of small birds frequent the nearby birchwood and the reedbeds, among them spotted flycatchers, tree pipits, redpolls, sedge warblers and reed buntings. In spring and autumn passage waders visit the shores, especially when water levels are low, and in winter whooper swans are among the wildfowl using the area. Wintering thrushes and finches regularly feed over the surrounding farmland, and birds of prey can sometimes be watched from the road as they hunt over Red Moss.

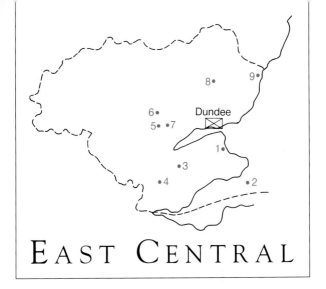

EAST CENTRAL

1 Eden Estuary LNR, St Andrews

Car parks at West Sands, St Andrews and Guardbridge on A91. Hide – key from North-east Fife Ranger Service (tel 0334 72151).

Because it is relatively narrow, with good vantage points along its southern shore, this estuary is easier than most to watch. The outer section is sandy; terns, often in large numbers, fish at the estuary mouth and the beaches are good for sanderling in autumn and winter. Shelduck feed over the muddy section all year and from late summer through to spring many waders and ducks are present. In winter seaducks, sawbilled ducks and divers can usually be seen offshore and sometimes within the estuary mouth.

2 Isle of May NNR, off Anstruther

Access by boat from Anstruther; day trips, weather and tide permitting, in summer. Visitor centre and summer warden. Bird Observatory (see Appendix).

Among the attractions of this low rocky island are its seabirds and breeding grey seals. It has the only large puffin colony on the east coast, as well as terns, gulls and auks and many nesting eiders. The cliffs are low, with many stacks and inlets, so it is possible to obtain good views of the cliff–nesting species. In spring and autumn the island is visited by a great variety of migrants; species such as goldcrest, warblers and thrushes sometimes arrive in large numbers with strong easterly or south-easterly winds.

3 Lomond Hills, Falkland

Access from car park on Falkland to Leslie road. Ranger service.

Upland birds breeding on this heather moorland area include red grouse, meadow pipit and cuckoo, with snipe in the boggy patches.

4 Loch Leven NNR/Vane Farm RSPB Reserve, Kinross

Access to Loch Leven NNR is permitted only at Kirkgate Park, Kinross, Burleigh Sands (north shore) and Findatie (south shore). Vane Farm has car parking, visitor centre (disabled access), hides, nature trail and warden service.

Loch Leven is one of Scotland's most important wildfowl lochs, with large numbers of breeding duck and wintering geese. Vane Farm includes wetlands behind the loch shore, open birch woodland and heather moorland, and consequently attracts a varied birdlife. From the observation room with its high-powered telescopes, or the main hide, ducks and waders around the man-made lagoons can be watched in summer, and feeding and flighting geese in winter. Tree pipits and spotted flycatchers are among the birchwood nesters, with redpolls and siskins often present in winter.

5 Hermitage & Craigvinean, Dunkeld

Signposted off A9 c1 mile west of Dunkeld.

The wooded riverside part of this area belongs to the NTS; the conifer woodland above is Forestry Commission plantation. A path leads from the Hermitage to Rumbling Bridge; dippers and grey wagtails breed along this stretch of river and many of the commoner woodland species are present among the trees along the route, with wood warblers around Rumbling Bridge. In addition to the usual birds of conifer plantations, the Craigvinean forest often has crossbills and jays and sometimes capercaillie.

6 Killiecrankie RSPB Reserve & Linn of Tummel, Pitlochry

Access from B8079 (old A9) Pitlochry to Killiecrankie. Car parks at Balrobbie Farm, Killiecrankie (RSPB Warden's house), Garry Bridge and Killiecrankie Visitor Centre.Ranger (NTS) & Warden (RSPB) services.

The east side of the wooded Pass of Killiecrankie belongs to the NTS; a network of paths and trails links at Garry Bridge with the Linn of Tummel path on the west bank, which leads on to Loch Faskally. The RSPB reserve stretches from the riverside oakwood to heather moorland above, and includes farmland and crags; unless accompanied by the Warden visitors should stay on the waymarked paths. Woodland birds in the Pass include wood warblers, pied flycatchers and both woodpeckers; kestrels, ravens and buzzards can often be seen around the crags; sawbilled ducks, dippers, grey wagtails and common sandpipers occur along the river; and the upland birds on the higher ground include black grouse.

7 Loch of the Lowes SWT Reserve, Dunkeld

Signposted off A923, 2 miles east of Dunkeld. Visitor centre (April–September), hide (disabled access), ranger service.

The ospreys breeding near the loch can be watched at the nest through the high-powered binoculars provided in the hide. Grebes and several species of duck breed on the loch or nearby, and in autumn are joined by roosting greylag geese and gulls and increased numbers of duck. The mixed woodland around the loch supports a good range of breeding birds, including redstarts and tree pipits, while redpolls and siskins are often present in winter.

8 Loch of Kinnordy RSPB Reserve, Kirriemuir

Access from B951 1 mile west of Kirriemuir. Car park, hides, summer warden.

This shallow marshy loch has a wider range of breeding grebes and ducks than most Scottish lochs: shoveler, gadwall, pochard, the American ruddy duck and black-necked grebe are among the less common species usually present. Black-headed gulls nest on the floating vegetation and water rails in the reedbeds; and scrub and woodland birds in the fringing trees and bushes. In autumn and winter grey geese roost on the loch and short-eared owls sometimes hunt over the marshes. Ospreys visit in July–August.

9 Montrose Basin LNR & SWT Reserve, Montrose

The most convenient access points are at the car park off the A935 on the western edge of Montrose, and at the roadside of the A92, near the southeast corner of the Basin. An SWT visitor centre is planned. Ranger service; hides on west shore require key obtainable from ranger (tel 0674 76336).

The good feeding on the extensive tidal mud and sand flats attracts large numbers of waders in autumn and winter; these are most easily watched when the tide is nearly in and the birds are driven to the edges of the Basin. Many wigeon and mallard are also present in winter, and several thousand pink-footed geese fly in to roost near the northeast corner. In summer shelduck, eider, oystercatcher and redshank are among the nesting waders, and a large flock of mute swans gathers to moult in July–August. At this time many terns also visit the Basin to fish.

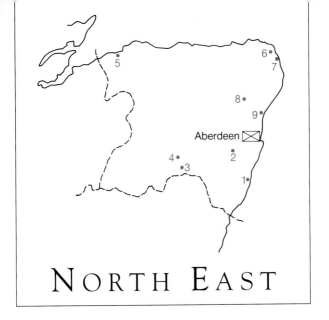

NORTH EAST

1 Fowlsheugh RSPB Reserve, Stonehaven

Signposted to Crawton off A92 3 miles south of Stonehaven. Car park.

Viewing of the large colony of seabirds, which includes some puffins, is most rewarding from May to July. Because the cliffs are deeply indented it is possible to obtain very good views of the birds at their nests without disturbing them.

2 Crathes & Drum, Deeside

Access off A93: Crathes 3 miles east of Banchory, Drum 3 miles west of Peterculter. Visitor centre and trails at Crathes, trails at Drum, ranger service (based at Crathes).

Both these NTS properties have extensive woodlands – there is ancient oakwood at Drum – with a wide range of breeding birds, including warblers, jay and both woodpeckers.

3 Glen Tanar NNR, Aboyne

Access via estate road off B976 at Bridge of Ess, c2 miles south west of Aboyne. Car park, visitor centre (April–September), summer ranger service, waymarked trails and start of long distance paths.

Much of this privately-owned estate is Scots pine forest, with extensive heather moorland on the higher ground to the south. Breeding birds of the forest include sparrowhawk, capercaillie, Scottish crossbill and siskin. Black grouse frequent the forest fringe and several species of predator hunt over the moorland; these upland birds are best seen from the Mounth and Firmounth 'Roads', the tracks of old drove roads, which lead over the hills to Glen Esk. Visitors should keep to the waymarked routes at all times and should not attempt to follow the tracks onto high ground unless properly equipped for the hills.

4 Muir of Dinnet NNR, Ballater

Access from B9119 (formerly A97), off A93 c4 miles east of Ballater. Car park beside B9119/A97 near Burn o' Vat.

The reserve comprises Lochs Davan and Kinord and a stretch of open birchwood and heather moorland lying to their west. Breeding birds include black grouse, and large populations of redpoll and woodcock in the woodland. In autumn and winter the lochs attract considerable numbers of duck and roosting geese, while merlins and hen harriers can often be seen hunting over the moorland.

5 Findhorn Bay & Culbin Forest, Moray

Findhorn Bay can be viewed from the B9011. Access to Culbin Forest is by minor road off A96 via Kintessack to Wellhill Farm or Cloddymoss.

Findhorn Bay is a good place to see fishing ospreys and terns. The commoner waders breeding around the Bay are joined by larger numbers in autumn and winter. Large gatherings of seaducks and divers are often present offshore from Findhorn village. Culbin Forest is one of the few planted pinewoods with breeding crested tits.

6 Loch of Strathbeg RSPB Reserve, Peterhead

Signposted off A952 at Crimond, c8 miles north of Peterhead. Visitor centre, hides, warden service.

The reserve covers not only the very large loch itself but also much of its surrounding marsh, farmland, dunes and woods. Breeding birds include shelduck and eider, water rail and terns, but the interest is greatest from autumn to spring. Large numbers of grey geese winter in the area and there are often several hundred whooper swans on the loch. Small parties of barnacle geese sometimes visit in autumn, and wintering duck include goldeneye, pochard and sawbills.

7 Rattray Head, Peterhead

Follow minor road to lighthouse, off A952 c1 mile south of Crimond.

This is one of the best places for watching the passage of seabirds, including skuas and shearwaters, in April–May and September–October. There are often divers offshore in winter.

8 Haddo House Country Park, Tarves

Signposted from Tarves on B999. Visitor centre, hides & ranger service.

The varied habitats within the park – both broadleaf and conifer woodland, open water and marshy ground, and parkland with scattered trees – provide suitable conditions for a good range of birds. Kestrel, sparrowhawk, great spotted woodpecker, goldcrest and several species of warbler are among the breeding birds. Ospreys occasionally visit the loch, which in winter attracts roosting greylag geese and a variety of ducks.

9 Forvie NNR, Newburgh

Access from the A975 just north of Newburgh. Car parks at (1) the bridge over the Ythan, from which paths lead north-eastwards across the moorland and along the riverside; (2) off the B9003 near Collieston, where there is a visitor centre; and (3) on the north bank of the river (approached by minor road off A975 opposite Collieston turning), where there is a hide. Warden service.

The reserve includes the narrow Ythan Estuary with its mudflats and mussel beds, low coastal moorland, a long sandy beach with extensive dunes behind, and low cliffs at the northern end near Collieston. In summer there are good numbers of breeding waders and shelduck and a large colony of eiders. Terns of four different species nest in a sanctuary area among the dunes near the river mouth; access is restricted from 1 April to 31 July but the birds can also be viewed from the south side of the river beyond the golf course. Seabirds breed on the cliffs at Collieston. From late summer to spring the estuary is used by a great variety of passage and wintering waders and wildfowl; the hide at (3) gives good views of the area where many of them congregate at high tide. Several thousand grey geese are often present in autumn and there are usually seaducks and divers off the rivermouth in winter.

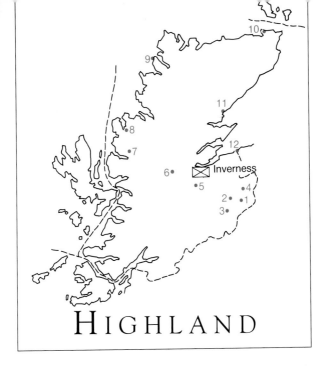

HIGHLAND

1 Cairngorm NNR, Rothiemurchus & Glenmore, Aviemore

Visitor centres at Inverdruie (Rothiemurchus Estate – map available), Glenmore Forest Park and Cairngorm Chairlift Car Park. Waymarked routes, nature trail at Loch an Eilein, ranger services.

Much of the privately-owned Rothiemurchus Estate lies within the vast Cairngorm NNR, which adjoins Glenmore Forest Park. The Rothiemurchus pinewoods are part of the ancient Caledonian forest; capercaillies, crested tits and crossbills are all present, though only crested tits are easy to locate along the many tracks. Ospreys can often be watched fishing at the Inverdruie fish farm, and a regular programme of guided walks is run by the estate rangers. Siskins, woodcock, black grouse and sparrowhawks are among the birds likely to be seen in Glenmore Forest Park, of which Loch Morlich is part, or in the small SWT reserve at the Pass of Ryvoan. The summit of Cairngorm can be reached via the chairlift, but visitors are warned to wear stout footwear and warm, windproof clothing. Ptarmigan can often be seen around the chairlift area, and dotterel breed on the plateau; snow buntings are scarce in summer but more numerous around the car park in winter. Golden eagles nest in the impressive mountain corries and may be spotted soaring high overhead.

2 Craigellachie NNR, Aviemore

Access from Aviemore Centre, via path under A9 from car park at dry ski slope.

The lower part of the reserve is birch woodland, in which great spotted woodpeckers, wood warblers, tree pipits, spotted flyctchers and redstarts are among the breeding birds. Above the woodland the ground rises steeply to heather moors; peregrines nest on the crag overlooking the Centre – look for the 'whitewash' of their droppings on the cliff face.

3 Insh Marshes RSPB Reserve, Kingussie

Access by B970 2¹/₂ miles from Kingussie. Car park, visitor centre, hides, trails, warden service.

Marshy ground covers much of the reserve, with some open pools, willow carr, rough grassland and birchwood on the higher and drier ground. An unusually wide range of wildfowl and waders breed in the marshes. Hen harriers and short-eared owls regularly hunt over the area, and in winter there are often up to 150 whooper swans present. Woodland nesting species include tree pipit and redstart.

4 Abernethy Forest – Loch Garten RSPB Reserve, Nethybridge

Signposted off B970 north of Aviemore. Hide (disabled access), warden service.

In addition to the well-known osprey nesting site, this part of Abernethy includes native pinewood, two lochs and boggy areas. Crossbills and crested tits inhabit the woodland, goldeneye may be seen on the lochs at any time of year, and in autumn Loch Garten has large roosting flocks of gulls, geese, ducks and sometimes whooper swans.

5 Loch Ruthven RSPB Reserve, Inverness

Signposted off B851 at East Croachy, 10 miles south of Inverness. Car park, hide (reached by 1/2 mile walk), summer warden.

The reserve is special for its slavonian grebes, which nest in the vegetation around the loch shore. Both common and black-headed gulls also breed, and red-throated divers occasionally come to fish. Redpolls and siskins are present in the nearby birchwoods, and black grouse at the moorland edge. Ravens are regularly seen in the area, and hen harriers and peregrines also frequently hunt there.

6 Glen Affric, Cannich

Access by minor road off A831 (Beauly to Drumnadrochit) at Cannich. Car parks at several points by the roadside.

Part of this typical highland glen is a Native Woodland reserve, with remnant old Caledonian pine forest. This is a good place to look for two pinewood specialities, crested tit and Scottish crossbill; they can often be seen along the paths leading from the car park at Dog Fall. Sawbilled ducks and occasionally black-throated divers occur on the lochs further up the glen, and golden eagles, buzzards and sparrowhawks are regular in the area.

7 Torridon, Kinlochewe

North of A896 9 miles southwest of Kinlochewe. Visitor centre at Torridon, near junction with minor road to Diabaig. Ranger naturalist.

This spectacular mountainous area is a NTS countryside

property. Paths lead into it from car parks on the A896 and the Diabaig road. Eagles can often be watched soaring high above the tracks, ptarmigan on the high ground and moorland birds at lower levels. Red-throated divers breed on some of the lochans. Ducks, waders and herons feed around the mouth of the River Torridon. The high hills should be visited only by those adequately equipped and experienced, or on a ranger-led guided walk.

8 Inverewe Garden, Poolewe

On A832 6 miles north of Gairloch. Car park, garden visitor centre, summer ranger-naturalist.

The main attraction of this NTS property is its magnificent garden, where exotic plants flourish thanks to the influence of the Gulf Stream, making it an oasis in an otherwise somewhat bare countryside. The lush growth attracts many small woodland birds, and this is one of the most likely places in the north west highlands to find wood warblers and chiffchaffs. Waders and eiders breed around the shore, and moorland birds on the rough ground nearby.

9 Handa Island SWT Reserve, west Sutherland

Access by boat from Tarbet, reached by minor road off A894 between Scourie and Laxford Bridge; daily sailings in summer (except Sundays), weather permitting. Summer warden.

The island has a huge seabird colony, with many good viewpoints from which the nesting birds can be watched. Great and arctic skuas, several wader species, wheatears and stonechats also breed on the island, and terns on the rocks near the landing place.

10 Dunnet Head, east of Thurso

Access by B855 off A836 c3 miles east of Castletown. Car park and good viewpoint near lighthouse.

The cliffs of Dunnet Head hold a good variety of breeding seabirds, including puffins and black guillemots, and also rock doves. Ravens, twites and a few great skuas are among the birds frequenting the moorland crossed by the approach road. In winter seaducks and divers are present in Dunnet Bay, to the west.

11 Loch Fleet SWT Reserve, Golspie

Access from minor road off A9 along south shore, to two car parks; or via minor road from Golspie past golf course to car park at Littleferry.

The reserve includes the tidal loch and some of the pinewood on the north shore. Shelducks, eiders, redshanks and oystercatchers are present throughout the year, and in summer both common and arctic terns – and occasionally ospreys – fish in the shallow water. Autumn sees a greater variety of ducks and waders visiting on their way south, and in winter seaducks and divers gather offshore. Numbers of long-tailed ducks are especially high in spring. The pinewoods hold crossbills and siskins, and in summer redstarts.

12 Culbin Sands RSPB Reserve, Nairn

Access, on foot only, is from Kingsteps, 1 mile east of Nairn on minor road past golf course.

A stretch of sandflats, saltmarsh and shingle bar, this reserve is of most interest in autumn, winter and spring, when large numbers of wildfowl and waders are usually present. The many hundreds of waders feeding over the flats and saltmarshes attract visiting birds of prey; peregrines, merlins and hen harriers can often be seen. Greylag geese roost on the shingle bar, and seaducks gather offshore. In summer there are breeding waders and a few terns.

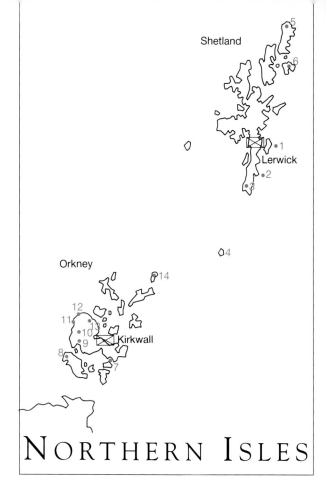

Shetland

•5

•6

♦

Lerwick •1

•2

♦3

♦4

Orkney

♦14

12

11 13

•10

•9

8

Kirkwall

♦7

NORTHERN ISLES

1 Noss NNR, Shetland

Access by ferry from Lerwick to Bressay, car or foot across the island, and small boat to Noss. An easier way to view the spectacular seabird cliffs is by an excursion boat trip from Lerwick Harbour.

Noss has one of the more accessible gannet colonies, with some 7,000 pairs breeding. There are also large numbers of other seabirds, including puffins, on the cliffs, and both arctic and great skuas on the moorland interior of the island.

2 Mousa, Shetland

Access by small boat from Leebotten, Sandwick (timetable from Tourist Office, Lerwick).

This small island has one of the few readily accessible storm petrel colonies; the birds can be both heard and smelled as one climbs up inside the walls of the ancient broch. Eiders, skuas, terns and black guillemots are among the other species nesting on the island, and there are usually seals on the shore.

3 Loch of Spiggie RSPB Reserve, Shetland

The loch can be viewed from the minor road which branches west from the B9122 at Scousburgh; one of the best viewing points is on the north shore, opposite the track to the beach. NO ACCESS TO THE SHORELINE.

In summer great and arctic skuas, kittiwakes and arctic terns regularly bathe in the loch, and teal, shelduck and waders breed around the shores. In spring long-tailed ducks gather before migrating, and in autumn whooper swans, greylag geese and several species of duck visit as they gradually move southwards.

4 Fair Isle, Shetland

Access by boat from Grutness, Shetland, or by air from Shetland. Permanently staffed Bird Observatory.

Fair Isle belongs to the NTS and has a thriving crofting community. The 17 seabird species breeding on the cliffs and moor include large numbers of puffins (easily watched and approached), many great and arctic skuas, and a recently established gannet colony. The island has long been known for the numbers and variety of migrant birds which arrive there in spring and autumn. The observatory was set up primarily for the study of migration but today it concentrates also on seabird studies. Accommodation is available both at Fair Isle Lodge, which is run in association with the observatory (see Appendix), and in some of the croft houses.

5 Hermaness NNR, Unst, Shetland

Access by car ferry from Yell or air from Lerwick, and B9086 road to Burrafirth, then on foot. Summer warden.

The best seabird cliffs on this large reserve are at the north end, reached by a well-defined path from the end of the road. Many thousands of gannets, shags, kittiwakes, fulmars and auks – including vast numbers of puffins – breed around the Hermaness coast, and skuas, dunlin, golden plover, snipe and twite are among the species nesting on the moorland interior. Red-throated divers breed on the small moorland lochans.

6 Fetlar (part RSPB Reserve), Shetland

Accessible by car ferry from Yell. Between April and September visitors to the RSPB reserve must be escorted by the summer warden; contact him at Bealance Croft, signposted 2$^1/_2$ miles from the ferry terminal.

Fetlar's cliffs and moorlands have a good variety of breeding seabirds, including terns and skuas, and on the headland of Lamb Hoga colonies of both storm petrels and manx shearwaters. The moorland area within the RSPB reserve, to the north of the road, is important for nesting whimbrels, and the small Loch of Funzie for red-necked phalaropes. Red-throated divers, ravens and twites also breed on the island.

7 Churchill Barriers, South Isles, Orkney

Some of the many seaducks, divers and grebes which winter in Scapa Flow can sometimes be watched from the A961 causeways which link Mainland with the islands of Burray and South Ronaldsay. A few birds usually stay until early May, by which time they are in breeding plumage and so easier to identify. There is also the possibility of seeing otters and seals in this area.

8 North Hoy RSPB Reserve, Orkney

Hoy can be reached by car ferry from Houton to Lyness or passenger ferry between Stromness and Moness. A minor road signposted to Rackwick runs through the RSPB's North Hoy reserve from the B9047, and a footpath starts about $1^1/2$ miles west of Moness at HY 223034 and leads round the west side of Ward Hill to Rackwick. From Rackwick a path leads to the Old Man of Hoy. There is a resident warden.

Large numbers of great and arctic skuas breed on the North Hoy hills, along with moorland species such as red grouse, golden plover and dunlin, stonechat and twite and a few hen harriers, merlins and short-eared owls. The cliffs hold a good range of nesting seabirds and also rock doves, ravens, peregrines and buzzards, and there are many pairs of red-throated divers on the hill lochans. A small patch of native deciduous trees at Berriedale supports a few woodland birds.

9 Lochs Harray & Stenness, Orkney Mainland

Viewing is possible from the B9055 roadside at a number of points, especially the car park for the Ring of Brogar, between the lochs.

The lochs are most interesting in autumn and winter, when they often hold several thousand wildfowl. Large numbers of pochards visit Harray, which is freshwater, while the tidal Loch of Stenness attracts goldeneye and long-tailed ducks. Whooper swans and greylag geese graze over farmland close to the lochs. Breeding species include the most northerly colony of mute swans, as well as tufted duck, mallard and waders around the shores.

10 The Loons RSPB Reserve, Orkney Mainland

Access to hide only, off minor road which leaves A986 3 miles north of Dounby.

A very wet marshy area, this reserve provides breeding habitat for several species of ducks and waders, including pintail, shoveler and snipe, and also arctic terns and small wetland birds such as sedge warbler and reed bunting.

11 Marwick Head RSPB Reserve, Orkney Mainland

Car parks at Cumlaquoy and Marwick Bay, both on minor roads off B9056.

The shelved cliffs along this stretch of coast are packed with guillemots and kittiwakes. Razorbills, puffins, ravens and rock doves are also present, in smaller numbers, and there are eiders and waders in and around Marwick Bay. There are a number of good viewpoints overlooking the breeding ledges. As at all cliff sites CARE MUST BE TAKEN WHEN NEAR THE EDGE OF THE CLIFFS.

12 Point of Buckquoy, Orkney Mainland

Car park and access causeway to Brough of Birsay not passable for 3 hours on either side of high tide.

This is a good place for seawatching, especially from July to October. Among the seabirds likely to be seen passing are manx shearwaters, storm petrels, skuas and terns.

13 Birsay Moors & Cottasgarth RSPB Reserve, Orkney Mainland

Access points and hides at (1) Cottasgarth, signposted to Lower Cottasgarth off minor road which leaves A966 3 miles north of Finstown, and (2) Burgar Hill, signposted off A966 at Evie.

Much of this large moorland reserve, which has breeding hen harriers and short-eared owls, can be viewed from the

B9057. The Burgar Hill hide overlooks a small lochan where red-throated divers can usually be seen in summer. Other breeding birds of the moorland include great and arctic skuas, merlins, golden plovers, dunlins and curlews.

14 North Ronaldsay, Orkney

Access by ferry or air, from Kirkwall.

North Ronaldsay has a surprising variety of breeding birds, which some years include such local species as pintail, shoveler, water rail and sandwich tern. There is a colony of cormorants, as well as gulls, fulmars and black guillemots. Like Fair Isle and the Isle of May, North Ronaldsay has become known for the 'falls' of migrants which occur in spring and autumn, and as a result the North Ronaldsay Bird Observatory was established in 1987 and now provides simple accommodation for visiting birdwatchers (see Appendix).

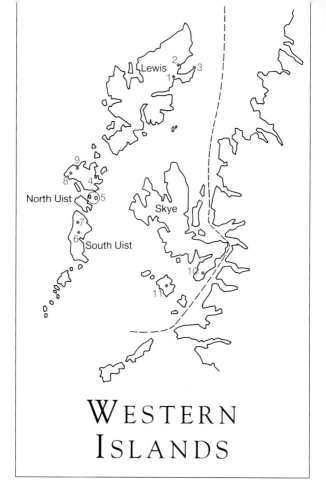

WESTERN ISLANDS

1 Stornoway Woods, Lewis

Nature trails.

This mixed wood, with its understorey of brambles and rhododendron is one of the very few places in the Outer Hebrides where a fair variety of woodland birds can be found. Among the species breeding there are treecreeper, great and blue tits, goldcrest, spotted flycatcher, rook and heron. A few pairs of ravens usually nest in the wood;

scavenging birds can be watched at the Stornoway rubbish tip, beside the A858 about 2 miles west of the town.

2 Gress to Tolsta, Lewis

The moorland around this stretch of the B895 north of Stornoway has breeding great and arctic skuas.

3 Tiumpan Head, Lewis

Approached by A866 east of Stornoway

This rocky headland is a good place for seawatching. It has some nesting seabirds, though not puffins, and usually also breeding ravens. Loch Branahuie, on the narrow neck between Stornoway and the Eye Peninsula often has ducks, especially in spring and autumn, while the nearby Melbost Sands, viewable from the little village of Steinish, provide feeding grounds for waders.

4 The A865/867 circuit, North Uist

When driving round this route it is worthwhile stopping occasionally to scan the moors and hills for hunting birds of prey, including golden eagles, and the many lochs for divers, both species of which breed in this area. Greylag geese also nest beside some of the lochs.

5 North Ford, between North Uist and Benbecula

The A865 causeway gives good views over the tidal flats to either side, where many waders feed at low tide. Eiders and shelduck are often present.

6 Loch Druidibeg NNR, South Uist

Good views can be obtained from the A865 and B890; access to the islands and south shore is restricted. Greylag geese and a variety of other wildfowl and of waders breed on the reserve, which includes moorland, machair and beach as well as the loch. Birds of prey which may be seen hunting over the surrounding moorland include golden eagles, buzzards and hen harriers. The small patch of mixed plantation beside the B890 is one of the few places in the Uists which support woodland birds.

7 Loch Bee, South Uist

Notable for its large gatherings of mute swans, sometimes several hundred strong, the loch can be viewed from the A865, which runs across it.

8 Balranald RSPB Reserve, North Uist

Access by minor road to Hougharry, signposted off A865 3 miles northwest of Bayhead. Summer warden, who should be contacted at reception cottage at Goular. NO ACCESS IN AND AROUND GRAVEYARD.

The reserve includes a rocky headland, sandy beaches backed by dunes, machair and marshland. Corncrakes breed regularly and a good variety of ducks nest around the marshes. The machair and croftland holds very large numbers of waders, including dunlin, and also corn buntings and twites. The headland at Aird an Runair is a good vantage point for seawatching in spring and autumn, when gannets, skuas, shearwaters, and auks are among the birds passing quite close inshore.

9 Vallay Strand, North Uist

The A865, which runs along the southern shore of the bay, offers good viewing of waders feeding over this great expanse of tidal mud and sand.

10 The Clan Donald Centre, Armadale, Skye

Access by boat from Mallaig or by A851 from Broadford. Visitor centre, waymarked walks, ranger service.

Armadale Castle, and the estate on which it stands, is owned by a Clan Trust. The estate includes planted and natural woodlands, hill ground and lochs, and consequently supports a good variety of birds. Nature trails give access to the amenity woodland. Rangers can provide information on other walks in the area.

11 Isle of Rum NNR

Access by ferry from Mallaig (also summer sailings from Arisaig). Nature trails, wardens.

The whole of the large and mountainous island of Rum is an NNR. Loch Scresort and the woodlands near Kinloch Castle are the most accessible areas, and can be visited during a day trip from the mainland. To go further afield it is necessary to stay on the island and accommodation must be booked in advance: camping, basic bothies, hostel and hotel available. Breeding birds include seabirds (Rum's manx shearwaters nest high on the hills but can often be seen offshore), golden and white-tailed eagles, other raptors and typical upland birds such as wheatear, stonechat and golden plover. Mature and recently-planted woodlands and the shoreline add habitat variety.

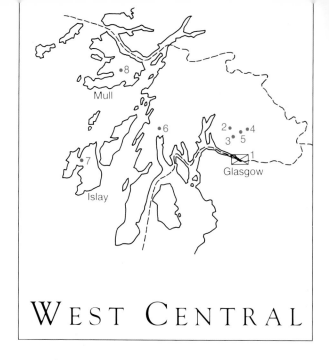

Mull

Islay

Glasgow

WEST CENTRAL

1 Possil Marsh SWT Reserve, Glasgow

Access at Lambhill Bridge over Forth & Clyde Canal. Parking in Skirsa Court, off A 879.

The different habitats in the reserve – shallow water, marsh, meadow and scrub – make it attractive to a good variety of birds. Water birds, warblers and reed buntings are among the breeding species, and in spring and autumn the area is an important staging post for migrants, especially waterfowl and waders.

2 Inversnaid RSPB Reserve, Loch Lomond

Access from Aberfoyle, via B829 and minor road (there is a postbus service), then on foot by the West Highland Way along the loch shore. Nature trail (steep), warden.

This RSPB reserve includes deciduous woodland and moorland on the higher ground above. It is notable for the pied flycatchers which breed there, many of them in nest boxes. Other woodland birds include great spotted woodpeckers, redstarts, wood warblers, tree pipits, woodcock and siskins. There are black grouse, buzzards and ravens on the moorland area, and sawbilled ducks on the loch.

3 Loch Lomond (part NNR) & Ben Lomond

Access to east shore of loch by B837 from Drymen; to Inchcailloch (part of NNR) by boat from Balmaha; to Ben Lomond by footpath from Rowardennan.

The woodlands along the east shore from Balmaha to Rowardennan, much of which lies within the Queen Elizabeth Forest Park, and on Inchcailloch, support a good variety of woodland birds, including redstarts, tree pipits, wood warblers and pied flycatchers. Conifer plantations cover large areas of the lower slopes of Ben Lomond, but the open ground above, which belongs to the NTS, has some of the commoner upland birds, such as red grouse, snipe and meadow pipit. Common sandpipers and grey wagtails are regular in summer along the loch shore.

4 Pass of Leny, Callander

Car parks at Kilmahog and Coireachrombie, on west side of river near south end of Loch Lubnaig. Ranger service.

The track along the south side of the river running through this narrow pass, in which are the Falls of Leny, passes through oakwood and alders, with conifers on the higher ground above. Among the woodland birds present in summer are wood and garden warblers, both flycatchers, siskins, redpolls and sparrowhawks. Goosanders, common sandpipers and dippers breed by the river, and there are moorland species, including ring ouzels and ravens, around the path from Coireachrombie to the summit of Ben Ledi (for this ascent appropriate footwear and protective clothing are necessary).

5 Queen Elizabeth Forest Park, Aberfoyle & Balmaha

Stretching from the Trossachs to Loch Lomond, this vast site includes lochs and rivers, moorland and hills, mixed deciduous woodland and conifer plantations. Visitor centre (David Marshall Lodge, on A821 $1/2$ mile north of Aberfoyle), nature trails. The western section of the Park is described under the entry for Loch Lomond.

The many routes, road and foot, through this area provide easy access to a wide range of habitats. Birds of the lochs and rivers include wigeon and goldeneye, grey wagtail, dipper and heron. Stonechats are often to be seen perched on gorse bushes near the Duke's Pass on the A821, and the remoter heathery areas have merlins and hen harriers as well as the commoner moorland species. Both woodpeckers are present in the deciduous woods of the Trossachs area and near Loch Ard, with flycatchers, tree pipits and wood warblers. The varied ages of the plantations provide suitable conditions for a good variety of species.

6 Argyll Forest Park

Access from A815 Dunoon–Strachur & A83 around Arrochar. Forest information centres at Ardgarten, Glenbranter and Kilmun.

A maze of tracks runs through this vast forest area, which stretches from Loch Lomondside to Strachur and south almost to Dunoon. The conifer plantations within the Park have breeding crossbills, goldcrests, siskins and a few capercaillies, with black grouse along the forest fringe. Above are extensive unplanted moorlands, with breeding red grouse, curlews, stonechats and twites, and over which several of the larger birds of prey hunt. Semi-natural woodlands on the shore of Loch Lomond, in Glen Loin and Glen Branter support a good variety of species, including jay and wood warbler, while the planted grounds of the Younger Botanic Gardens at Benmore and the Kilmun Arboretum are likely spots for blackcap, garden warbler and chiffchaff.

7 Islay (part RSPB Reserve)

By ferry from Kennacraig, west Loch Tarbet, or air from Glasgow. Loch Gruinart RSPB Reserve, visitor centre, hide & warden.

The Loch Gruinart reserve is best known for its wintering barnacle geese, which usually start to arrive in late September. In summer there are wildfowl and waders breeding on the wetter low-lying ground, while black and red grouse and hunting raptors of several species can be seen on the moorland. Other good areas to visit are (1) the Rhinns: the crofting ground north of Portnahaven to hear corncrakes in summer, and the Rubha na Faing headland for seawatching in April/May and August–October; (2) the Oa, for choughs, ravens and breeding seabirds; and (3) the inner end of Loch Indaal, around Bridgend and Bowmore, especially in winter, for waders and wildfowl, including scaup and roosting barnacle geese.

8 Mull

Access by ferry from Oban, Lochaline or Ardnamurchan

Although there are no reserves or country parks on Mull there are many good opportunities for birdwatching. Breeding birds of prey include golden eagle, hen harrier, peregrine and merlin, which may be seen hunting over the extensive moorlands and young plantations in Glen More (A849) or around the Mishnish Lochs (B8073 near Tobermory). Both whinchats and stonechats are present, and red-throated divers often fish on the Mishnish Lochs. Around the coast in summer there are common and arctic terns, red-breasted mergansers, eiders and the commoner shoreline waders. Gannets and other seabirds can often be seen offshore.

SOUTH WEST

1 Baron's Haugh RSPB Reserve, Motherwell

1/2 mile from Motherwell town centre, signposted from Adele Street, opposite Civic Centre. Hides (disabled access), nature trail, warden.

Lying close to the River Clyde, the reserve includes marshland, meadows, scrub and woodland. It supports a good variety of wetland species, with residents including kingfisher and water rail as well as various ducks. In autumn waders on passage south visit the muddy marshland, and in winter whooper swan, shoveler, pochard and goldeneye are often present. Viewing can be especially good in frosty weather, as one area of the marsh generally remains unfrozen and consequently attracts many waterfowl close to a hide. Sand martins nest in the river banks, and other summer visitors to the area include whinchat and grasshopper warbler.

2 Falls of Clyde SWT Reserve, New Lanark

Visitor centre, nature trail, ranger service.

The stretch of river within the reserve is the haunt of dippers, grey wagtails and occasionally kingfishers. Kestrels nest on the steep side of the gorge, and breeding birds of the mixed woodland include willow tits and four other species of tit, garden warblers and both woodpeckers.

3 Caerlaverock NNR & Eastpark, Dumfries

Signposted off B725 between Glencaple and Bankend. Visitor centre and hides at Eastpark. Warden service: guided tours at Eastpark 11am and 2pm daily, mid–September to late April.

The NNR covers a large stretch of the foreshore at Caerlaverock; Eastpark is a Wildfowl and Wetlands Trust Refuge, where the farmland is managed for the benefit of wildfowl. The area is an important wintering ground for barnacle geese, which can be most easily watched from the Eastpark hides. Pink-footed and greylag geese also visit the farmland, and the pools at Eastpark attract whooper swans and good numbers of ducks, including wigeon, pintail and shoveler. Many waders feed over the mudflats and roost at high tide on the saltmarsh, and birds of prey regularly hunt over the area.

4 Galloway Forest Park, New Galloway/Newton Stewart

The main access points are along the A712 and off the A714 to Glen Trool. A forest drive runs from Clatteringshaws on the A712 to the A762 beside Loch Ken. Trails.

Extending over a vast area, the Park includes moorland and hills, lochs and rivers, conifer plantations, scrub and deciduous woodland. At Glen Trool the oak and birch woodlands have breeding redstarts, pied flycatchers, wood warblers and tree pipits, and to the east willow tits frequent the damp woods near Loch Ken and nightjars breed in the forest nearby. Peregrines, hen harriers, merlins and short-eared owls hunt over the moorlands within the park, there are a few ravens on the hills and black grouse along the woodland fringe, and siskins and crossbills nest in the plantations.

5 Wood of Cree RSPB Reserve, Newton Stewart

4 miles from Newton Stewart, beside minor road from Minigaff along east side of River Cree, parallel to A714. Warden, marked trails.

In summer the deciduous woodland, which contains many oaks, has nesting wood warblers, tree pipits, pied flycatchers and redstarts. Woodcock, sparrowhawk, buzzard and willow tit are among the resident species, and there are breeding barn owls in the area.

6 Mull of Galloway RSPB Reserve, Stranraer

22 miles south of Stranraer: A716 to Drummore, then B7041 and minor road to lighthouse.

Seabirds breeding on the reserve include cormorant and black guillemot as well as most of the cliff-nesting species. A few puffins are usually present and gannets from the colony on the nearby Scare Rocks can be watched fishing off the headland. Corn buntings nest on the farmland behind the Mull.

7 Culzean Country Park, near Ayr

Signposted off A77 at Maybole. Visitor centre, ranger service.

Much of this NTS property is wooded, but ponds, streams and a stretch of shoreline also lie within the park. Breeding species include a good range of woodland birds, especially warblers, and shelduck where the shore is sandy. Fulmars nest on the Castle cliffs, and oystercatcher and redshank are also present in summer. Many wildfowl winter on Swan Pond, and divers and seaducks can be seen offshore.

8 Lochwinnoch RSPB Reserve

Access from A760 Largs–Paisley road. Visitor centre, nature trails, hides, warden.

The reserve comprises two wetland areas – the open Barr Loch and marshy Aird Meadows – and a patch of adjoining mixed woodland. Wildfowl breeding on the reserve include shoveler, mallard and teal, and several pairs of great crested grebes nest. In winter wildfowl numbers and variety increase, with whooper swans, greylag geese, goosanders and goldeneye usually present, and often a few cormorants. Grasshopper warblers are among the small birds breeding in the reedbeds and woodland.

9 Brodick Country Park & Goatfell, Arran

Ferry from Ardrossan. Visitor centre, nature trail & ranger service at Brodick Castle.

This extensive area is owned by the NTS. The mixed woodlands of the Brodick country park and the neighbouring Merkland Wood support a good variety of breeding birds, including several warbler species. Nightjars nest in the area. Moorland birds breed on the lower slopes of Goatfell, while ravens, ptarmigan and golden eagles are regularly seen higher up.

GOING FURTHER WITH BIRDWATCHING

Anyone who becomes really interested in birds will soon want to know much more than can be included in this small volume – about how to identify the rarer species, where and when to visit to have the best chance of seeing particular species, and the populations of each species present in different parts of Scotland. No single book provides all this information, but those suggested below cover all three aspects. It is also helpful to make contact with other birdwatchers, which can be done through the Scottish Ornithologists' Club and the Royal Society for the Protection of Birds (addresses below). And for those who want to handle, as well as watch, birds, the best thing to do is either visit a Bird Observatory, or contact a local Ringing Group.

IDENTIFICATION BOOKS
There are many field guides dealing with all the birds of Britain and Ireland; among the best are:

Collins Field Guide to Birds of Britain and Europe by R.Peterson, G.Mountfort and P.A.D.Hollom, published by HarperCollins.

The Shell Guide to the Birds of Britain and Ireland by J.Ferguson-Lees, I.Willis and J.T.R.Sharrock, published by Michael Joseph, illustrates seasonal plumages of British species and also includes descriptions of species which occur irregularly as vagrants.

BOOKS ON PLACES TO VISIT AND SPECIES DISTRIBUTION
Where to Watch Birds in Scotland by M.Madders and J.Welstead, published by Christopher Helm, gives details of

many good birding locations, with notes on access, best times to visit, and the species likely to be seen.

A Guide to the Nature Reserves of Scotland, published by Macmillan, and the RSPB's leaflet *Nature reserves – information for visitors* also provide useful information on sites.

Birds in Scotland by V.M.Thom, published by Poyser, describes the status and distribution of Scottish birds, and the changes which have taken place over the last few decades.

Many areas have their own local bird reports, which often include information on birdwatching sites. A list of these is available from the Scottish Ornithologists' Club (see below).

ORGANISATIONS CONCERNED WITH BIRDWATCHING IN SCOTLAND

The Scottish Ornithologists' Club, 21 Regent Terrace, Edinburgh EH7 5BT (tel 031 556 6042). The Club has 14 Branches around Scotland which have programmes of winter meetings and summer outings, and runs two conferences every year. It publishes a quarterly newsletter *Scottish Bird News*, a twice-yearly journal *Scottish Birds* and an annual *Scottish Bird Report*.

The Royal Society for the Protection of Birds – Scottish Headquarters, 17 Regent Terrace, Edinburgh EH7 5BN (tel 031 557 3136). The RSPB has many reserves in Scotland; a leaflet is available with details of visiting arrangements. Local Members' Groups arrange meetings, outings and fund-raising events, and branches of the Young Ornithologists' Club cater for young people up to 18 years old.

Fair Isle Bird Observatory, Fair Isle, Lerwick, Shetland ZE2 9JU (tel 035 12 258) is permanently staffed and carries out important studies on seabirds and migrants. Fair Isle is renowned for the numbers of rare birds which land there during peak migration periods. Fair Isle Lodge, run in conjunction with the Observatory, provides excellent accommodation with full board from late April until the end of October.

The Isle of May Bird Observatory (in the Firth of Forth) provides good opportunities for watching both seabirds and migrants. Basic accommodation (self-catering) is available from mid–March to the end of October. Further details can be obtained from: R.Cowper, 9 Oxgangs Road, Edinburgh EH10 7BG.

North Ronaldsay Bird Observatory, Orkney, offers simple self-catering accommodation for up to 20. Details from Dr K.Woodbridge, Twingness, North Ronaldsay, Orkney.

Ringing Groups are active in several parts of the country, trapping a wide range of species by a variety of methods. By accompanying a qualified ringer those wishing to handle live wild birds can obtain the experience necessary before a licence to do so is issued. Information on Ringing Group contacts can be obtained from *The British Trust for Ornithology*, The Nunnery, Thetford, Norfolk IP24 2PU (tel 0842 750050).

The Scottish Wildlife Trust, Cramond House, Cramond Glebe Road, Edinburgh EH4 6NS is concerned not only with birds and their habitats but with all Scottish wildlife. The Trust's reserves include a number with special bird interest.

INDEX

THE PRINCIPAL REFERENCE TO EACH SPECIES/GROUP IS
SHOWN IN **BOLD**